Street Drainage Design and Modeling

STREET DRAINAGE DESIGN AND MODELING

By

James C.Y. Guo, PhD, P.E. Professor

Water Resources Publications, LLC Since 1971 WRP Colorado

For Information and Correspondence:
Water Resources Publications, LLC
P. O. Box 630026, Highlands Ranch, Colorado 80163-0026, USA

STREET DRAINAGE DESIGN AND MODELING

By

James C.Y. Guo, PhD, P.E. Professor

ISBN-13: 978-1-887201-91-9

Library of Congress Control Number: 2017948829

This publication is printed and bound in the United States of America.

Foreword

Urban drainage system is one of the major components in the urban infrastructure systems. Stormwater imposes potential hazards to moving vehicles. A quick removal of storm runoff from streets, roads, and highways is critically important to maintain the level of services in transportation systems. Although the primary function of a street is to facilitate traffic movement, during a storm event streets serve as wide and shallow waterways to collect and to remove stormwater from urban areas. The objective of street stormwater drainage design is to warrant efficient and safe traffic movements during the severe storm events. This book is dedicated to the latest developments in: (a) new concepts to analyze the urban catchment hydrology for storm runoff predictions, (b) innovative methods to estimate the street allowable capacities to convey storm runoff, and (c) useful computer models to simulate flow movements in inlets and sewers.

The new concepts include the revision of the Rational method. Although the Rational method is the oldest conventional approach for urban hydrology studies, in this book, the *Rational method* was revised to quantify the flow increases due to urban development. A new procedure was adopted to reflect the impact of urban imperviousness on the time of concentration. The more the pavements are developed in the catchment, the shorter the overflow length will be in the flow movement. As a result, the time of concentration needs an adjustment for the future development potential. The revised Rational method is an imperviousness-based approach to reflect the impact of urbanization on the increase of peak flow.

The conventional approach recommended in the design manuals of US Federal Highways does not take safety and debris clogging into considerations when estimating an inlet flow interception. In this book, it is recommended that the flow capacity in a street be evaluated by the gutter-full flow capacity with a discharge reduction. In addition, the decay-based clogging factor was presented in this book for single to multiple inlets. As a recommended in the new approach, the clogging factor shall be applied to the length for weir flows or to the area for orifice flows.

Sewers in a system are first sized individually under the assumption of normal flow, and then evaluated by the hydraulic and energy grade lines. The latest computer model of UDSEWER is introduced in this book as an efficient tool to design a sewer system using the mixed flows among open channel flow, surcharge flow, and pressure flow in a system. It calculates hydraulic and energy grade lines with consideration of friction loss, bend loss, jump loss, and manhole loss. On top of flow hydraulics, UDSEWER summarizes the excavated earth volumes for the entire sewer system for future cost estimation.

Photos and figures are efficiently used to illustrate how and why the engineering approaches were derived from and applied to street drainage designs. Many design procedures presented in this book have been converted into Excel Spread Sheets. All these efficient tools can be downloaded from the website: www.udfcd.org at no cost.

This book is the summary of the author's 30-year research publications and class notes developed for graduate classes and senior design courses. More than 100 real world design examples are used to illustrate the methods and procedures. Many of the design methods documented in this book have been adopted as the recommended design procedures by Denver, Las Vegas, and Sacramento metropolitan areas in the United States. For additional information, the reader can visit the websites: www.udfcd.org.

James C.Y. Guo, PhD, P.E. Professor/Director
University of Colorado Denver, Denver, Colorado

Acknowledgements

The preparation of this book was supported by the Urban Watersheds Research Institute (UWRI) and the author is grateful to the Institute and its Board of Directors. Mr. Ben Urbonas, President of UWRI, and Mr. Ken MacKenzie, Director of Urban Drainage and Flood Control District (UDFCD), Denver, Colorado provided comments and reviewed chapters and sections of this book. Many methods and design procedures have been adopted and translated into the Urban Stormwater Design Criteria Manuals supported by UDFCD.

The Urban Drainage and Flood Control District (UDFCD) in Denver Colorado was the source of field data, photos, and maps that supplemented the examples presented in this book. The website, www.udfcd.org provides the software and Excel Spreadsheets/Books that all readers and users of the UDFCD's methods can access.

Table of Contents

CHAPTER 1
STREET DRAINAGE SYSTEM

1.1. DRAINAGE FACILTIES

Urban drainage system is one of the major components in the urban infrastructures. A quick removal of storm runoff is critically important to maintain the level of service of roads and highways in an urban area. The primary function of a street is to facilitate traffic movement. However, during a storm event, stormwater imposes potential hazards to moving vehicles. The objective of street stormwater drainage design is to warrant efficient and safe traffic movements during the severe storm events. A drainage system is designed to *collect* stormwater from the pavement surfaces, *convey* it through gutters and sewers, and *discharge* it to a downstream receiving water body without causing on- and off-site adverse impacts. As illustrated in Figure 1.1, a street drainage system is composed of the following major elements as:

(1) *Collection facilities* ---- inlets, culverts,

(2) *Conveyance facilities* ---- gutters, sewers, and

(3) *Storage facilities* --- water quality or flood control basin.

Inlets and sewers along a street are often sized to collect and to convey the minor storm event, i.e. 2- to 5-year storm. As illustrated in Figure 1.1, street inlets drain into manholes, and sewers are the links between manholes. The outfall sewer drains into the downstream waterway. When the storm sewer system is overwhelmed by a large storm event, the excess storm water will have to be spread out on the streets. After the year of 2000, stormwater management in the metropolitan areas is subject to the mandates of water quality control under the *Federal Clean Water Act* (CWA 1972). A storm sewer system with its outfall sewer<36 inch in diameter must drain into a water quality enhancement facility to remove solids and pollutants. As illustrated in Figure 1.2, the water quality basin is designed to have a resident time of 12 to 24 hours. Usually 70 to 80 percent of solids carried in urban stormwater are expected to be trapped in the basin. A major sewer system with its outfall sewer≥36 inches in diameter needs a detention basin system that is designed to provide both flow release and water quality controls (Guo 1999). The ultimate goal is to release no more than the pre-development peak flows during the extreme events (10 to 100-yr storm events).

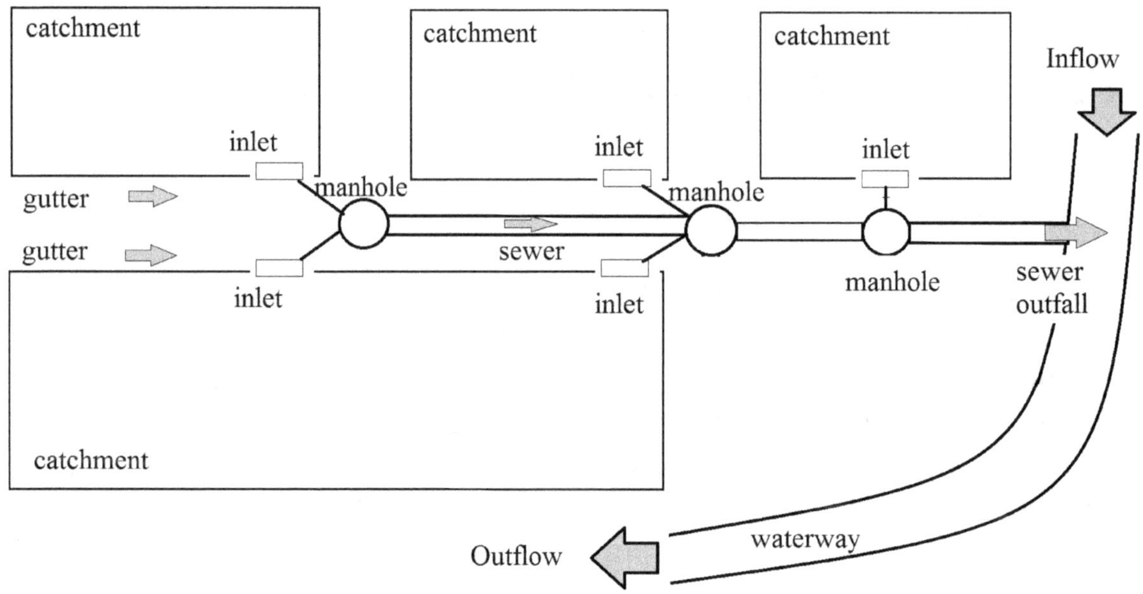

Figure 1.1. Storm Sewer System.

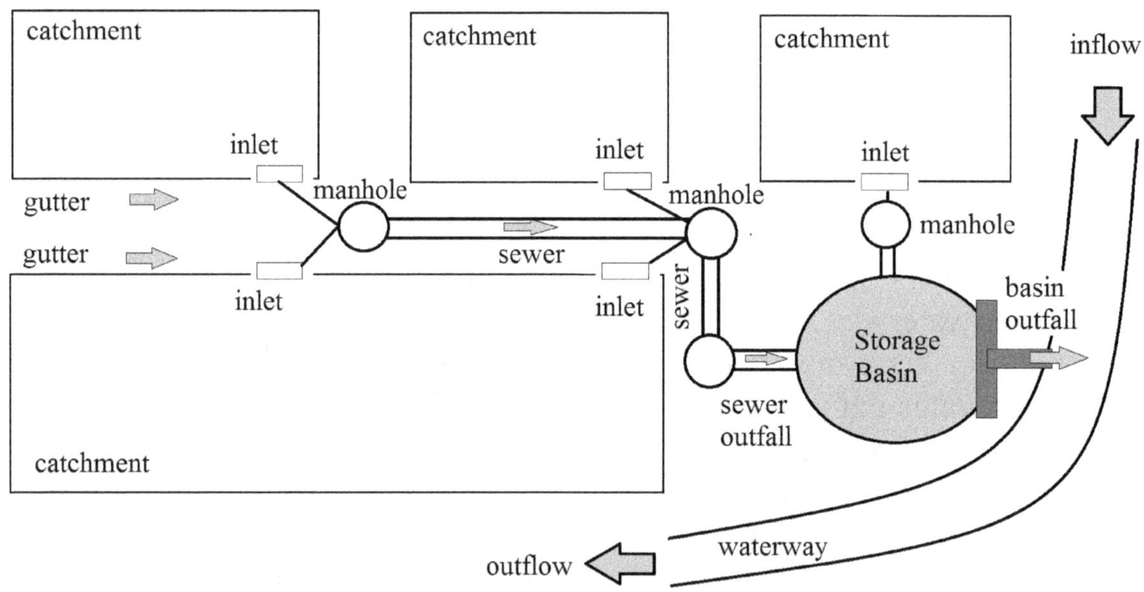

Figure 1.2. Storm Sewer-Storage System.

At the outfall point, it is necessary to have energy dissipaters to provide a smooth transition from water to be released from the sewer outlet into the basin. The basin also has its outfall system to release flow into the downstream water body.

1.2. MINOR AND MAJOR SYSTEMS

Storm sewer systems are designed to collect the minor flood flows from the streets. A minor event is defined by its return period. For instance, the peak flow of a 2-yr event is expected, on an average, to be exceeded or equaled once every 2 years. Based on engineering economics, the 2 or 5-yr event is often recommended for street inlet and storm sewer designs. During the major storm event such as the 10- to 100-year events, the excess storm runoff will overflow into the adjacent streets. The flood flow depth along a street will temporally surcharge manholes and sewers until the flood flow recedes.

Before the development in an urban area, it is important to conduct a *regional storm water drainage plan* for the entire watershed. Urban developments are staged over a period of time. A site development needs to have a *local stormwater drainage plan* conducted to make sure that the flow releases from the site agree with the allowable flow release defined in the regional drainage plan. The objectives of a storm water drainage plan can only be achieved if the designer understands that the operations of the *dual drainage systems* coexist in an urban area. Storm sewers provide the conveyance capacity for smaller but frequent events while street gutters provide the channel flow system for extreme events. Street drainage is a joint operation of the *Minor and Major Drainage Systems*. During a major event, a street operates like a wide and shallow channel that carries the storm runoff in excess of the capacity of the storm sewers.

EXAMPLE 1.1 The inlet in Figure 1.3 is designed to collect the minor event of 10 cfs from the street. During the 100-year event, the peak runoff is estimated to be 25 cfs immediately upstream of the inlet. It is expected that the inlet intercepts 10 cfs and then leaves a residual flow of 15 cfs into the street gutter.

This simple approach of flow diversion did not take the effect of hydraulic head into consideration. In reality, a slightly more flow can be carried by the sewers when the street gutter flow becomes deeper than the curb height during a major storm.

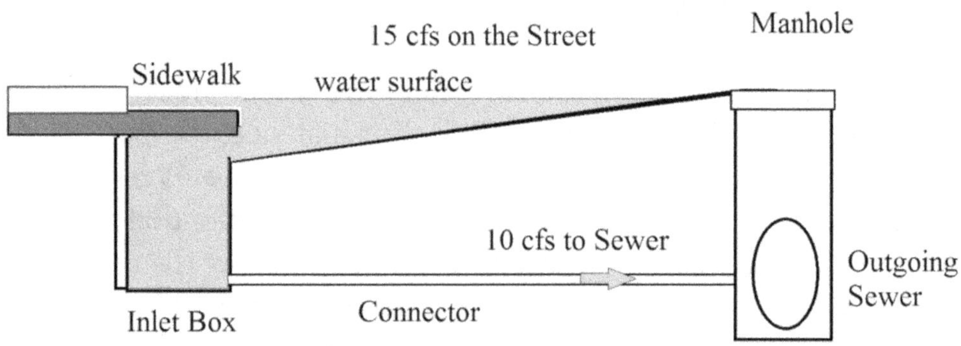

Figure 1.3. Illustration of Dual Drainage Systems.

1.3. STREET CLASSIFICATIONS

Based on the average daily traffic volume, streets are classified into *local, collector, arterial, and freeway* systems. Details are discussed as follows:

a) *Local Streets*

A local street is built with two traffic lanes, stop signs and roadside parking spaces. The width of a local street is 30 to 36 feet.

b) C*ollector Streets*

Local streets are merged into a collector street that has stop signs at the street intersection. A collector street may have a width of 36 to 44 feet for two to four moving traffic lanes with car parking adjacent to the curbs.

c) *Arterial Streets*

Arterial Streets are further classified into major arterial streets and minor arterial streets. They may have traffic signals at major intersections. They may have four to six traffic lanes. Parking adjacent to the curbs is prohibited.

d) *Freeways*

Freeways are designed for rapid and efficient movement of traffic. Access is controlled with ramps at an interchange. A freeway may use as many as eight lanes. Car parking along the curbs of a freeway is prohibited.

Water on a street imposes hydroplaning effects that can severely impact on the traffic system to a level of paralysis. To remove storm runoff from the traffic lane areas, a street cross section in Figure 1.4 is designed to have a street crown at the center of the street width. A transverse slope of 2% extends from the street crown to the curb. At the curb face, a gutter depression of 2 inches is introduced to increase the hydraulic capacity in the gutter. Street gutters can be prefabricated with a gutter width of 2 feet and a curb height of 6 inches (HEC12, 1984).

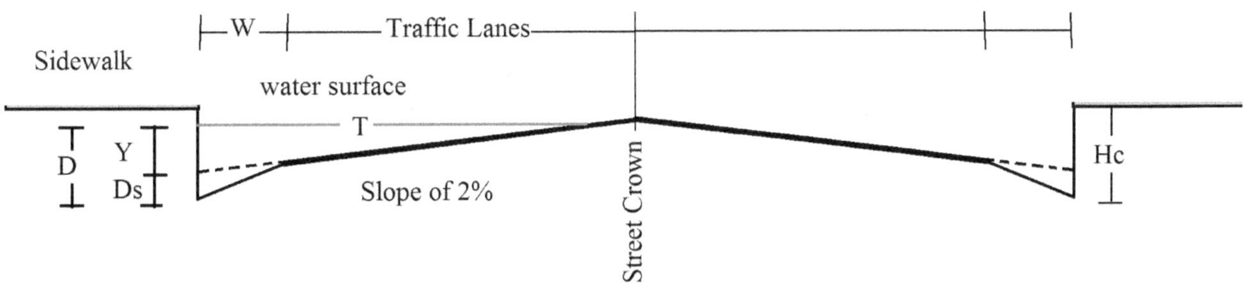

Figure 1.4. Typical Urban Street Cross Section.

$$D = Y + D_s \tag{1.1}$$

$$T = \frac{Y}{S_x} = \frac{D - D_s}{S_x} \tag{1.2}$$

Where D= flow depth in gutter in [L], D_s= gutter depression in [L] such as 2 inches, T= water spread in [L], S_x = transverse or side slope in [L/L], W= gutter width in [L] such as 2 feet, and H_c= curb height in [L] such as 6 inches. Care must be taken on units when operating Eq's 1.1 and 1.2.

It is important to keep in mind that the primary function of a street is for traffic movement. The drainage function is considered secondary. Therefore, storm water design must not interfere with the traffic functions and safety. For instance, any transverse slope greater than 2% would compromise driver's comfort. The water spread on the street is sensitive to its transverse slope. A shallow sheet flow on the street can be hazards to the traffic safety. A proper drainage system is manifested by effectively collecting and conveying water flows into sewers or roadside ditches (HEC22, 2010).

1.4. RECOMMENDED DESIGN CRITERIA

The goal of street drainage design is to direct storm runoff towards street gutters. The water spread on the street must be maintained within the allowable width that is determined by the traffic safety and the needs of emergency traffic. During a storm event, the top width (or water spread) of the water flow widens as more stormwater is collected into the gutter. The spread of water would eventually hinder traffic flow and possibly become hazardous to moving vehicles. Based on these considerations, the design criterion for street drainage designs were developed using the maximum allowable water spread or gutter flow depth as the standard. As illustrated in Figure 1.5, under a minor event, the sidewalk is not allowed to be overtopped by the gutter flow in a local street. It implies that the maximum allowable gutter flow depth in a local street is not exceeding the curb height. For a collector street, an additional design constraint is to keep at least one traffic lane free from water due to the emergency needs. Under a major storm event, the gutter flow depth is not exceeding 18 inches for both local and collector streets.

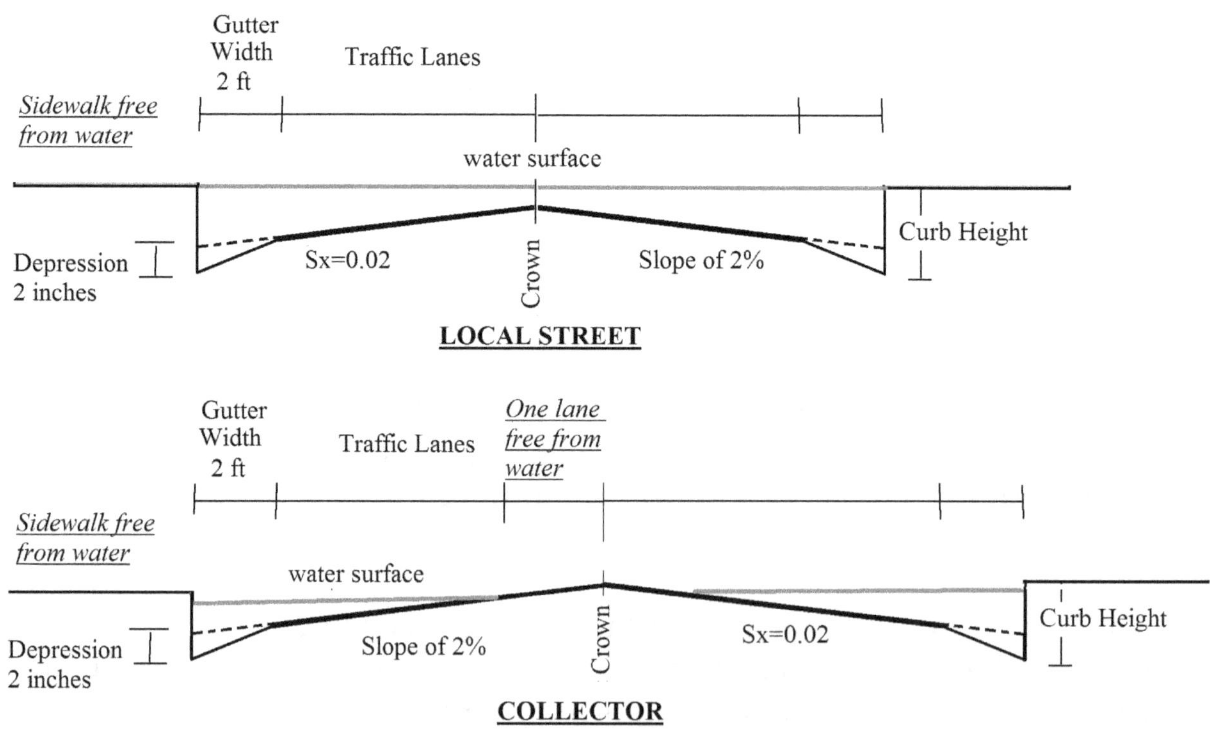

Figure 1.5. Design Criteria for Street Drainage under Minor Event.

The gutter flow in Figure 1.6(a) did not overtop the sidewalk; therefore it was under a minor event. On the contrary, the event in Figure 1.6(b) was under a major storm because the gutter flow depth exceeded the curb height.

(a) Minor Event (b) Major Event

Figure 1.6. Minor and Major Flows.

EXAMPLE 1.2 A local street has a width of 30 feet with the standard street cross section as shown in Figure 1.7. A half street width is divided into a gutter width of 2 feet and traffic lanes of 13 feet. According to the design criteria under a minor event, the sidewalk must be free from the gutter flow. Verify that the street width dictates the maximum allowable flow depth in the gutter.

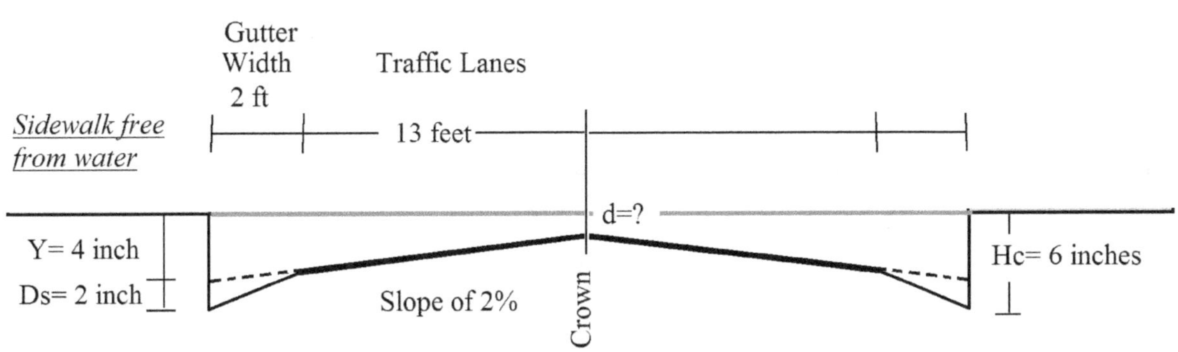

Figure 1.7. Flow Depth in Local Street.

$$S_x = \frac{(Y-d)}{T} = \frac{(4-d)/12}{13} = 0.02$$

d=0.88 inch

EXAMPLE 1.3 A collector street has a width of 44 feet with the standard street cross section as shown in Figure 1.8. A half street width is divided into a gutter width of 2 feet, traffic lanes of 13 feet, and emergency lane of 5 feet. According to the design criteria, the sidewalk must be free from the gutter flow under a minor event. Verify that the curb height dictates the maximum allowable flow depth in the gutter.

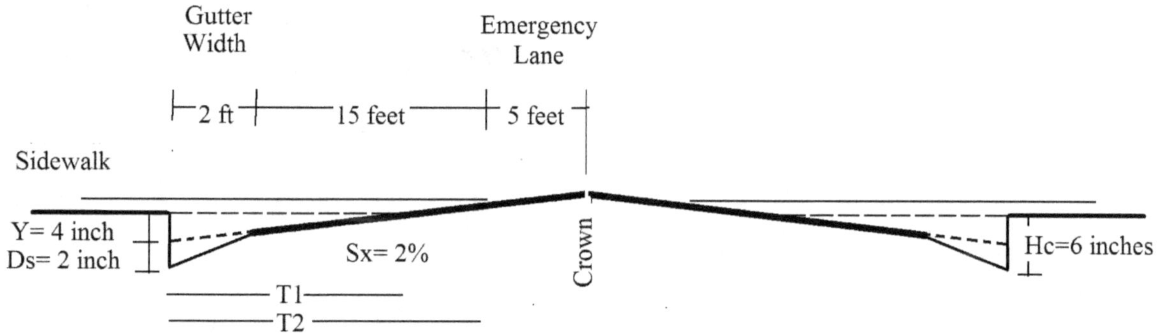

Figure 1.8. Flow Width in Collector Street.

Solution:

Case A. Accept the width of 17 feet. Let T_2= 17 feet.
The corresponding water depth, D, in the street gutter is:

$$D = Y + D_s = T_2 S_x \times 12 + 2 = 17 \times 0.02 \times 12 + 2 = 6.08 \textit{ inches} > 6 \textit{ inches} \quad \text{(No!)}$$

It results in a submerged sidewalk. Case A is not acceptable.

Case B. Accept the gutter full depth of 6 inches. Let D=Hc= 6 inches
The corresponding water spread, T_1, is calculated as:

$$T_1 = \frac{Y}{S_x} = \frac{(6-2)/12}{0.02} = 16.7 \textit{ feet} < 17 \textit{ feet} \quad \text{(Yes!)}$$

It results in an emergency lane slightly wider than 5 feet. Case B is acceptable.

1.5. CLOSING

Urban stormwater collection and conveyance systems are one of the major components in all urban infrastructure systems. It is essential to design these systems so carefully that flood damage and traffic impediment in urban areas can be minimized during storm events. A proposed street drainage system must follow the street layout. Therefore, both the design and construction of street inlets and sewers demand close coordination and cooperation between hydraulic and transportation engineers. During the different phases of construction, the minimization of traffic detour and the reduction on traffic delay also impose additional constraints on the sewer and inlet designs (CDOT Manual 2015). It is important to know that the optimal street drainage design is to maximize the street storage capacity. In other words, we shall delay the first inlet as farther downstream as possible because the first inlet begins to demand the manholes and sewers to convey the collected storm runoff.

1.6. HOMEWORK

Q 1.1 The gutter flow depth varies with respect to the storm event. The 100-yr gutter flow depth as shown in Figure HW1-1 is 42 inches while a depth of 36 inches is estimated for the 5-yr gutter flow. Considering that the orifice formula is applicable to estimate the flow released from the inlet box into the connector pipe of 15 inches in diameter, determine the difference between the 100-yr flow, *Q100*, and the 5-yr flow, *Q5*, released into the sewer line.

Hint: The *orifice formula* states: $Q = C_d \frac{\pi d^2}{4}\sqrt{2g(h-d/2)}$ in which Q= flow rate in *cfs*, C_d= discharge coefficient of 0.60, g = gravitational acceleration as 32.2 feet/second^2, h= headwater depth in feet, and d= pipe diameter in feet.

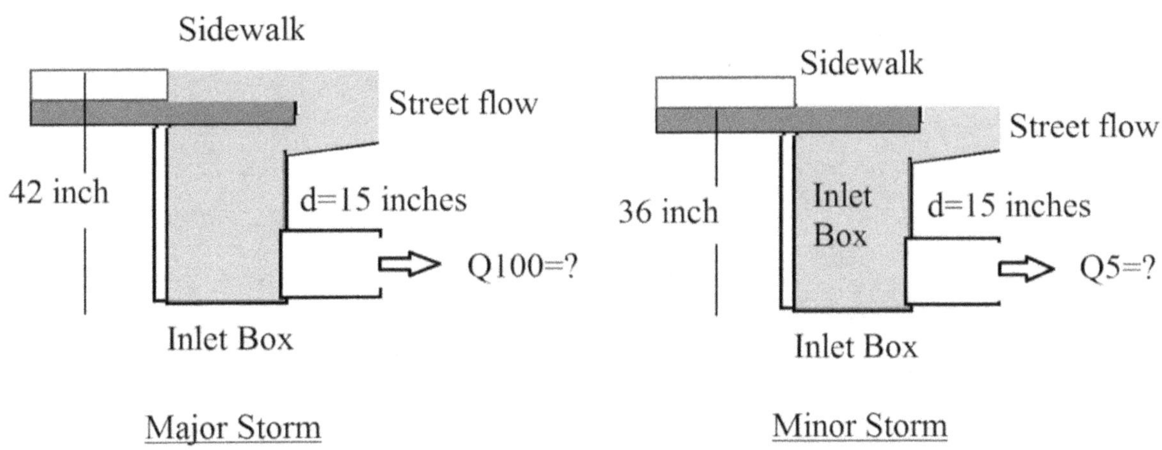

Figure HW 1-1. Flow Difference between Minor and Major Events.

Q1.2 Repeat EXAMPLE 1.3 with $S_x = 0.01$. Determine the gutter flow depth and water spread in EXAMPLE 1.3.

1.7. REFERENCES

CWA (1972). https://www.epa.gov/laws-regulations/history-clean-water-act

CDOH Manual (2015). "Hydraulic Design Manual", Colorado Department of Transportation.

Guo, James C.Y. (1999). "*Design of Detention Basins for Small Urban Catchments.*" ASCE J. of Water Resources Planning and Management, Vol. 122, No. 1., November.

HEC12 (1984). "*Drainage of Urban Highway Drainage".* Hydraulic Engineering Circular No. 12, US Department of Transportation, Federal Highways Administration.

HEC22 (2010). "Urban *Drainage Manual",* Hydraulic Engineering Circular No. 22, US Department of Transportation, Federal Highways Administration.

CHAPTER 2
PRECIPITATION AND RAINFALL STATISTICS

2.1. HYDROLOGIC CYCLE

Hydrologic cycle is the circulation of water through the atmosphere, lands, lakes, and oceans. Figure 2.1 indicates that the hydrologic cycle involves precipitation, surface runoff, groundwater, and evaporation. Precipitation takes many different forms for the moisture in the air to condense, to lift up, to fall, and to reach the ground. During this process, raindrops experience the interception by trees and buildings before reaching the ground. On the ground, raindrops first infiltrate into the soils and then fill up potholes and depressed areas. The infiltrating water will either percolate to reach the groundwater table or become a subsurface runoff moving toward the nearby streams and rivers. During a dry season, the groundwater table supplies a base flow to rivers and to lakes as well (Chow al. et. 1988).

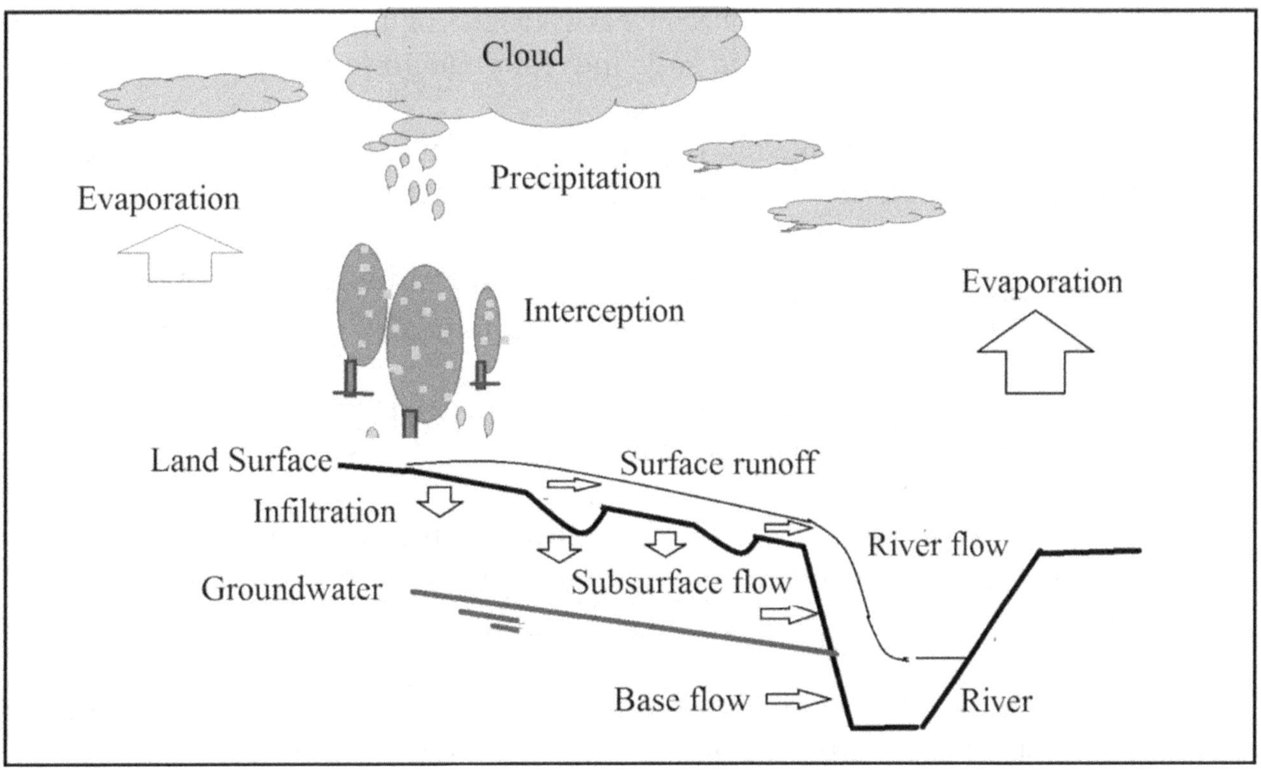

Figure 2.1. Hydrologic Cycle.

The rainfall excess is the amount of precipitation that survives the hydrologic losses and produces overland runoff toward streams, creeks, rivers, lakes, and oceans. Oceans and lakes are major sources for evaporation which is the process to convert the liquid water on the ground into moisture back to the atmosphere.

2.2. HYDROLOGIC LOSSES

Interception loss in a rainfall event is the amount of precipitation intercepted by trees and buildings. *Surface depression loss* is the rainfall amount entrained in the depression areas. For convenience, hydrologic losses are expressed in the same unit as the rainfall depth which is expressed as water depth per unit surface area. It is important to recognize that "depth per watershed" is a unit of volume.

In an urban area, the interception loss ranges between 0.1 and 0.2 inch. The depression losses are recommended to be 0.4 inch on pervious areas and 0.10 inch on impervious areas.

Interception and depression losses are local. However, soil infiltrations apply to the entire pervious surfaces in the watershed. Based on the infiltration characteristics, the US *Natural Resources Conservancy Services* (NRCS) or formally U. S. *Soil Conservation Service* (SCS 1976) has classified soils into four categories: Types A, B, C, and D.

Type A Soil ---- Soils having high infiltration rates between 0.30 and 0.45 inch/hr if thoroughly wetted and consisting mainly sand and gravel.

Type B Soil ---- Soils having moderate infiltration rates between 0.15 and 0.30 inch/hr if thoroughly wetted and consisting mainly of loamy soils.

Type C Soil ---- Soils having slow infiltration rates between 0.05 and 0.15 inch/hr if thoroughly wetted and consisting mainly fine textures.

Type D Soil ---- Soils having very slow infiltration rates between 0.01 and 0.05 inch/hr if thoroughly wetted and consisting mainly of clay soils with a swelling potential.

The water depth infiltrating into the soils per unit time measures the soil infiltration rate. The common unit for *infiltration rate* is inch/hour or mm/hour. The *soil infiltration amount* is the total water depth in [L] seeped into the soils for a specified period of time. As illustrated in Figure 2.2, at the beginning of a rainfall event, the soil medium is often so dry that the initial infiltration rate is very high. As time goes on, the soil medium becomes saturated and the infiltration rate decays to a constant rate. There are many formulas developed to describe this decay nature. Among them, the Horton's formula has been widely accepted and recommended for design. The *Horton's formula* states:

$$f(T) = f_c + (f_0 - f_c)e^{-KT} \tag{2.1}$$

in which $f(T)$ = infiltration rate in [L/T] at the elapsed time T, f_c = final (constant) infiltration rate in [L/T], f_o = initial infiltration rate in [L/T], and k = decay coefficient in [1/T] that has a unit of 1/second, 1/minute, or 1/hour, depending on the time unit used in Eq 2.1.

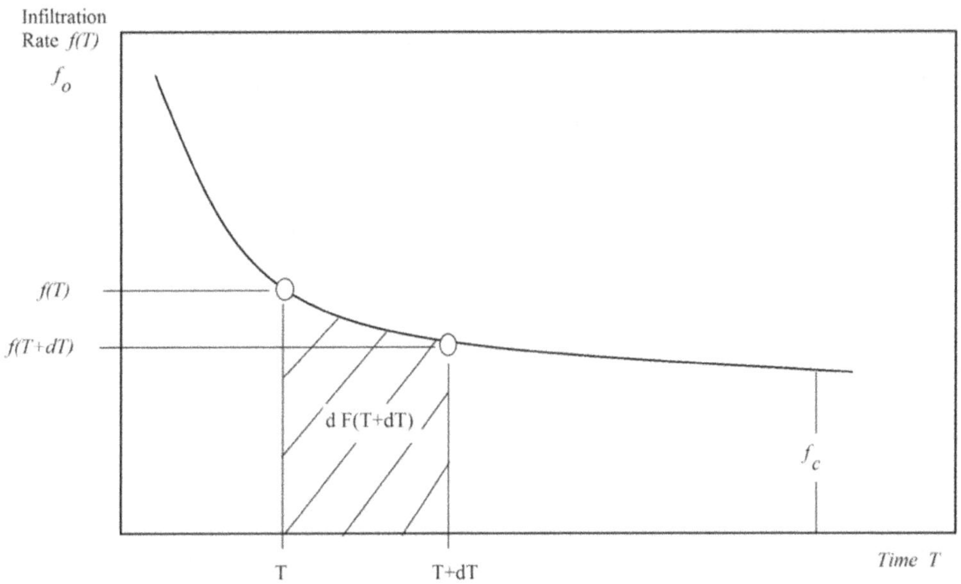

Figure 2.2. Variation of Soil Infiltration Rate with Respect to Time.

Table 2.1 presents the recommended infiltration rates for SCS four hydrologic types of soils.

Table 2.1. Recommended Soil Infiltration Rates (UDFCD 2010).

Soil Type	Initial Rate (inch/hr)	Final Rate (inch/hr)	Decay Coeff (1/hr)
A	4.50	0.60	6.48
B	4.00	0.55	6.48
C	3.00	0.50	6.48
D	3.00	0.50	6.48

The unit for cumulative infiltration amount is inch or mm per watershed. The infiltration amount is the area under the curve of infiltration rate. Mathematically, the infiltration amount, $\Delta F(T)$, can be calculated by the arithmetic average infiltration rate as:

$$\Delta F(T+dT) = 0.5\,[f(T) + f(T+dT)]\,dT \tag{2.2}$$

in which $\Delta F(T+\Delta T)$ = incremental infiltration in [L] from T to $T+\Delta T$, and ΔT = incremental time in [T]. Through a rainfall event, the cumulative infiltration amount, $F(T)$, up to time, T, is then calculated as:

$$F(T) = \sum_{t=0}^{t=T} \Delta F(t) \tag{2.3}$$

The hydrologic loss in an event is the sum of interception loss, depression loss, and infiltration loss. Subtracting the hydrologic losses from the total rainfall depth results in *rainfall excess* or *net rainfall*. Often the net rainfall depth is referred to as *runoff depth*.

$$R = P - I_v - D_v - F \tag{2.4}$$

$$C = \frac{R}{P} \tag{2.5}$$

in which R = runoff depth in [L], P = total precipitation depth in [L], L_v = interception depth in [L], D_v= depression depth in [L], F = soil infiltration depth in [L], and C= runoff coefficient representing the percentage of rainfall amount that is converted into surface runoff. The total runoff volume in $[L^3]$ from the event is the product of the runoff depth multiplied by the watershed area.

EXAMPLE 2.1 A rainfall event has a precipitation depth of 2.6 inches over duration of 60 minutes. Considering the interception loss of 0.2 inch, depression loss of 0.4 inch, and soil infiltration loss with f_c = 0.5 inch/hr, f_o = 3.0 inch/hr, k = 0.0018 per second, and ΔT = 5 minutes, determine the runoff depth for this 60-minute event.

Solution

Applying Eq's 2.1, 2.2 and 2.3, the infiltration rates and cumulative amounts are calculated in Table 2.2. The total infiltration amount for this event is 0.92 inch.

Table 2.2. Calculations of Infiltration Rates and Amount.

T minutes	$f(T)$ inch/hr	$\Delta F(T)$ inch	$F(T)$ inch
0	3		0
10	1.35	0.36	0.36
20	0.79	0.18	0.54
30	0.60	0.12	0.66
40	0.53	0.09	0.75
50	0.51	0.09	0.84
60	0.50	0.09	0.92

According to Eq 2.4, the runoff depth is then calculated as:

$$R = 2.6 - 0.2 - 0.4 - 0.92 = 1.08 \text{ inch.}$$

$$C = 1.08/2.6 = 0.415$$

For this case, the hydrologic loss is 58.5% of the total precipitation, and only 41.5% of the precipitation becomes surface runoff.

2.3. RAIN GAGE

Rainfall depths are recorded at a rain gage. The tipping bucket is a popular apparatus used for rainfall measurements. As illustrated in Figure 2.3, a tipping bucket gage is composed of an *orifice* to receive raindrops, a *bucket* to store the rain amount captured, and a *recorder* to register the incremental rainfall depths according to the clock time. The raindrops intercepted by the orifice opening are first funneled into the bucket. The cumulative water depth in the bucket triggers the electronic pen to continuously register the accumulation of rainfall depth on a roll of paper loaded on the revolving drum, which rotates according to the clock time. The bucket has a limited storage volume. The bucket can tip and pour the accumulated water into the reservoir as soon as the bucket becomes full. From the rainfall record, the distributions of cumulative and incremental rainfall depths can be plotted according to the clock time.

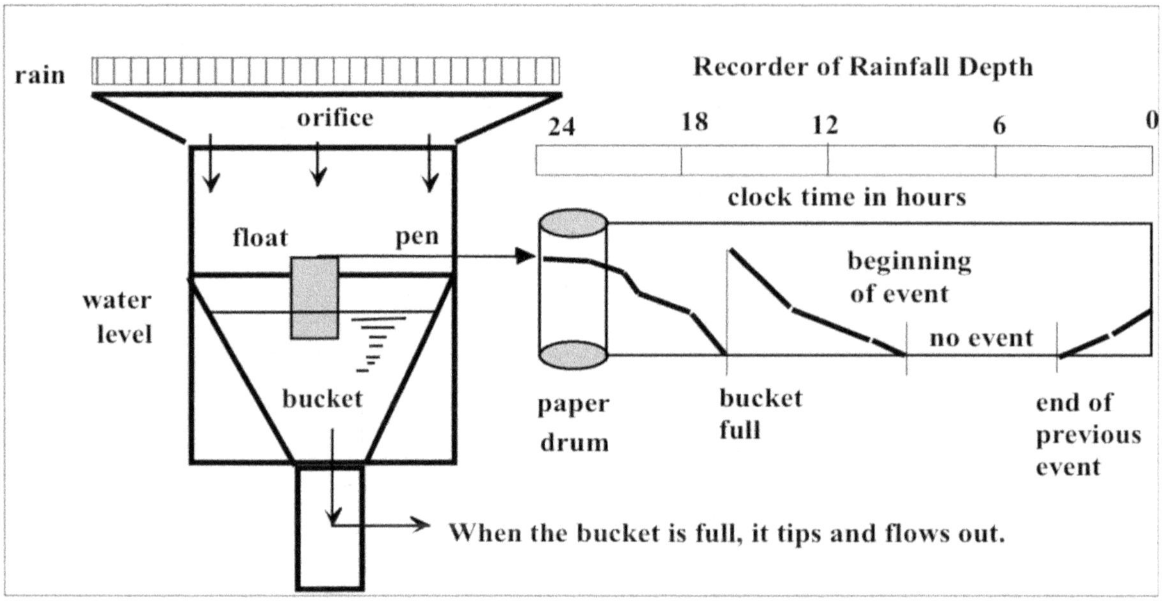

Figure 2.3. Tipping Bucket Rain Gage.

(a) Weather Station	(b) Rain Gage

Figure 2.4. Rain Gage in Field.

2.4. RAINFALL TIME-DISTRIBUTION

Rainfall characteristics are local and seasonal. During an event, the rainfall intensity and amount are varied temporally and spatially. A rain gage is installed to record the "point" measurement. The analyses of a rainfall event include the following major parameters:

a. Event Depth, P,
b. Event Duration, D,
c. Rainfall Depth Time-Distribution
d. Event Average Intensity, I

$$I = P/D \tag{2.6}$$

$$P(T) = \sum_{T=0}^{T=D} dp(T) \tag{2.7}$$

Where I = rainfall intensity in [L/T], P= rainfall depth in [L], D=duration in [T], Δp(T)= incremental depth in [L] between time T and T+ΔT, ΔT= time increment in [T]. Although the rainfall is a continuous process, the incremental rainfall depths are reported as a discrete record at a pace of one reading every 5-, 15-, or 60 minutes. The readings are often expressed in terms of cumulative rainfall depth. A *rainfall depth distribution* is the plot of incremental or cumulative rainfall depth with respect to clock time. A rainfall depth time-distribution is the history of the event and often used in stormwater numerical simulations.

2.5. RAINFALL DURATION-DISTRIBUTION

Having the rainfall depth distribution constructed, the event can be further analyzed by various hydrologic methods. *Duration analysis* is a procedure to identify the highest intensity for a specified time period. Figure 2.5 presents a 30-minute rainfall distribution that consists of six 5-minute rainfall incremental depths or six rain blocks for simplicity. Consider a period of five minutes as duration. This event can be separated into six 5-minute rain blocks. Each block is considered as a 5-minute independent event with its own depth and intensity. Similarly, consider a10-minute duration. Any two consecutive 5-min rainfall blocks constitute an independent 10-minute rain block. By following this concept, a 30-min rainfall distribution can be bisected into five 10-minute rain blocks, four 15-minute rain blocks, so on and so forth (Guo et. al. 2001).

The main purpose of duration analysis is to select the highest precipitation amount for a specified duration. For instance, the highest among the six 5-minute rain blocks is chosen to represent the 5-minute rainfall depth for the recorded event. Similarly, the highest among the five 10-minute rain blocks is chosen to represent the 10-minute rainfall depth for the recorded event. In doing so, a set of rainfall *depth-duration* (P-D) pairs can be derived from the rainfall depth distribution.

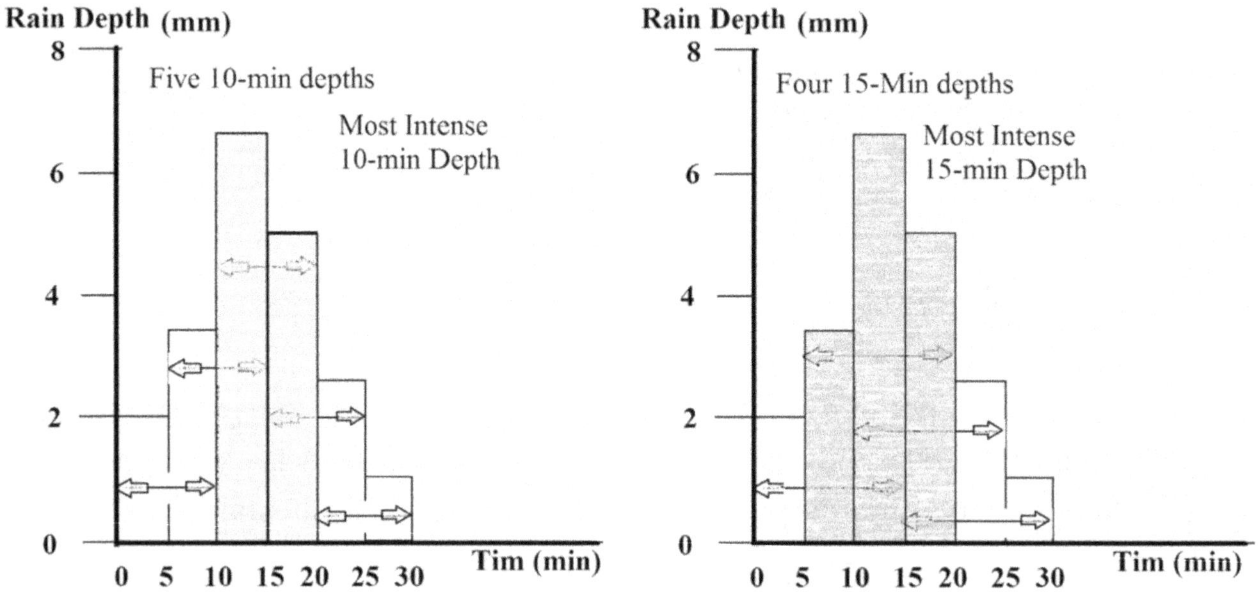

Figure 2.5. Illustration of 10- and 15-min Rainfall Depths.

The ratio of *rainfall depth* to *duration* is termed *rainfall intensity*. The plot of rainfall depth (P) versus duration (D) is termed *P-D curve* and the plot of rainfall intensity (I) versus duration (D) is termed *I-D curve*. A P-D curve is an increasing function of time, starting from the highest 5-minute depth to the total depth for the entire event. On the contrary, an I-D curve is a decreasing curve with respect to time, starting with the highest 5-minute intensity to the event average intensity.

<u>EXAMPLE 2.2</u> Table 2.3 presents an rainfall record observed from 16:00 to 17:00 . The incremental rainfall depths are registered in the 2nd column. The event rainfall depth was 0.79 inch over a period of 60 minutes or an intensity of 0.79 inch/hr. For this case, the P-I and I-D Curves are plotted in Figure 2.6.

Table 2.3. Rainfall Depth-Time and Depth-Duration Distributions.

Clock time T hr:min	Rainfall Increment Δ p(T) inches	Cumulative Rain Depth P(T) Inches	Duration D min	Depth p(D) inch	Intensity I(D) inch/hr
14:30	0.01	0.01	5.00	0.22	2.64
14:35	0.02	0.03	10.00	0.40	2.40
14:40	0.04	0.07	15.00	0.53	2.12
14:45	0.13	0.20	20.00	0.63	1.89
14:50	0.22	0.42	25.00	0.67	1.61
14:55	0.18	0.60	30.00	0.71	1.42
15:00	0.10	0.70	35.00	0.73	1.25
15:05	0.04	0.74	40.00	0.75	1.13
15:10	0.02	0.76	45.00	0.76	1.01
15:15	0.01	0.77	50.00	0.77	0.92
15:20	0.01	0.78	55.00	0.78	0.85
15:25	0.01	0.79	60.00	0.79	0.79

Rainfall Event Duration D 60.00 minutes

Rainfall Event Depth P 0.79 inch

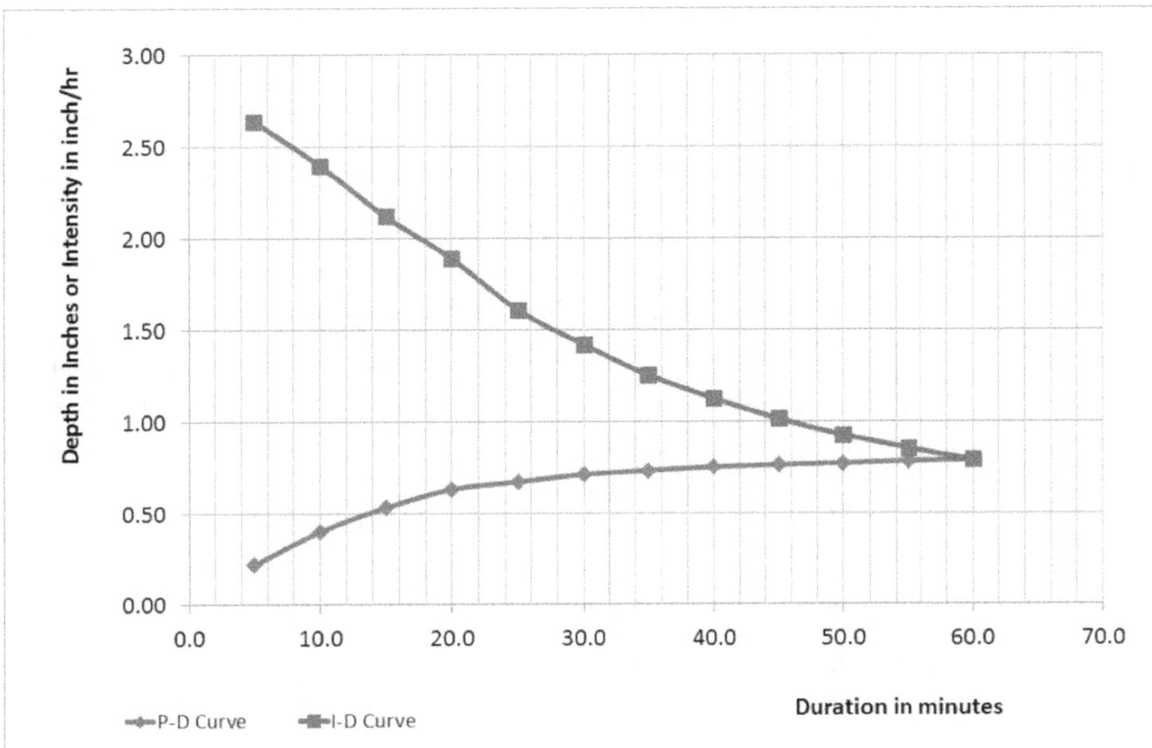

Figure 2.6. P-D and I-D Curves.

2.6. HYDROLOGIC FREQUENCY ANALYSIS

Hydrologic events are random in nature. Understanding of the local hydrology depends on the analysis of the local hydrologic data collected over a long period of time. Hydrologic data are recorded as a continuous time series. It takes a record of 10 years or longer in order to produce a reliable statistical analysis (Beard 1962).

Frequency analysis is a statistical method developed to determine the probability of outcome using the annual database such as *annual maximum series* (AMS). To construct an AMS, the continuous record is divided into segments. For instance, a water year is defined as a period of 365-day from October 1 to September 31 or from May 1 to April 30n depending on the local rainfall-snowfall characteristics. Using a water year as a basis, a long-term continuous record can be separated into 365-day segments. From a 365-day segment, the highest observed value is selected to represent the water year. For instance, the rainfall record observed at the Stapleton Airport gage station indicated that there were 5 storm events recorded in 1981. The highest 10-min value from each storm is selected to form a data base of five 10-min rainfall depths for the year of 1981. Among these five values, the highest one is chosen to represent the year of 1981. In doing so, a continuous record for 30 years will offer 30 data points for an AMS. In other words, the number of observations in an AMS is equal to the number of water years in the record. The statistical implication of an AMS is revealed from a ranking process among the data points. Details are explained by the following example (Bulletin 17, 1982).

EXAMPLE 2.3 Table 2.4 presents a 10-year record of 6-hour rainfall depths. The frequency analysis is tabulated in Table 2.4

Table 2.4. Example for Frequency Analysis.

Year	Depth inches		Rank	Ranked Depth inches	Approximate Return Period Years
1980	6.00		1	9.50	10.0
1981	9.50		2	9.00	5.00
1982	4.00		3	8.00	3.30
1983	8.00		4	8.00	2.50
1984	5.00		5	7.50	2.00
1985	9.00		6	6.75	1.70
1986	7.50		7	6.25	1.40
1987	6.25		8	6.00	1.25
1988	6.75		9	5.00	1.11
1989	8.00		10	4.00	1.00

The length of the record for this example is 10 years. During the period of 10 years, the highest value of 9.50 inches was equaled once. Therefore, the return period to exceed or equal the magnitude of 9.50 inches is approximately 10 years. Statistically, the 10-year 6-hour rainfall depth at the gage site is 9.50 inches. The second highest value of 9.00 inch in Example 2.3 was exceeded once and equaled once, or a total of twice in a period of 10 years. On an average, the return period to exceed or equal the magnitude of 9.0 inches is 5 years. Therefore, statistically, the 5-yr 6-hr rainfall depth at this rain gage is 9.0 inches.

2.7. STATISTICAL APPROACH

For an AMS, the return period of a magnitude can be determined by the empirical formula as (HEC SSP 2005, Peak Q Program 2005):

$$T_r = (N+1)/m \tag{2.8}$$

in which T_r = return period in years, N = number of years in the record, and m = rank of magnitude. For instance, when N=10, the rank for P = 9.0 inches in Example 2.3 is m = 2.0. Therefore, the return period for 6-hour precipitation≥9.0 inches is:

$$T_r = (10 + 1)/2 = 5.5 \text{ years}$$

It implies that the chance to exceed or equal 9.0 inch for a 6-hour event is, on average, once every 5.5 years. The non-exceedance probability is defined as the chance to have a rainfall depth, D, less than the referenced magnitude, d. Mathematically, such a probability is defined as:

$$P(D<d) = 1 - 1/T_r \tag{2.9}$$

Consider the above example. The non-exceedance probability for having a 6-hour rainfall depth less than 9.0 inch is

$$P(D<9.0) = 1 - 1/5.5 = 0.82$$

There are many statistical models developed for frequency analysis (Han 1977). The basic technique is to plot the observed magnitudes versus non-exceedance probabilities on a selected statistical graphical paper. If the statistical model can fit the observed data to the best, the trend of data on the graphical paper shall provide reliable predictions (Gringorten 1965, Guo 1986).

2.8. RAINFALL-INTENSITY-FREQUENCY (IDF) CURVE

The rainfall distribution varies with respect to time and space. The rainfall pattern is also local and seasonal. For instance, the rainfall characteristics in the front range of the Rocky Mountains are highly influenced by the orographic condition. In winters, the rainfall event tends to be longer than six hours and the amount of precipitation is relatively high, but its intensity is low. In general, a winter storm does not impose immediate flood threats to small urban catchments, but it can induce a huge flood after a long duration on a large watershed. On the other hand, a thunderstorm is often formed by a convective type of precipitation. The typical pattern of thunderstorms is characterized by short duration with a high intensity. Often, a thunderstorm imposes a higher chance of flooding.

For design, the selection of rainfall duration depends on the size of the watershed. The larger the watershed, the longer the duration. For urban runoff studies, the sizes of urban catchments are usually so small that a 2-hr rainfall distribution is recommended. For instance, the 1- and 2-hr rainfall depths are recommended for drainage studies in the metro Denver area. The one-hour precipitation depths, $P_{1\text{-hr}}$, for the Denver metropolitan area are presented in Table 2.5 as (UDFCD 2010):

Table 2..5 One-hr Precipitation Depths for Denver, Colorado.

Return Period in Years	2	5	10	50	100
One-hr Precipitation Depth in inches	0.95	1.35	1.61	2.21	2.61

According to the study of rainfall characteristics in front range of Rocky Mountains by the *National Atmosphere and Ocean Administration* (NAOA), the rainfall depth for any duration less than one hour can be linearly correlated to the one-hour rainfall depth by the empirical formulas as (Atlas 2 in 1976):

$$P_{30-minute} = 0.79 P_{1-hr} \tag{2.10}$$

$$P_{15-minute} = 0.57 P_{1-hr} \tag{2.11}$$

$$P_{10-minute} = 0.45 P_{1-hr} \tag{2.12}$$

$$P_{5-minute} = 0.29 P_{1-hr} \tag{2.13}$$

in which $P_{30\text{-minute}}$= rainfall depth in inches for duration of 30 minutes, and so on. Equations 2.10 to 2.13 form a data base for regression analyses to develop the rainfall *intensity-duration-frequency* (IDF) formula which can be modeled by a hyperbolical formula. For instance, using the Denver rainfall depths as an example, the best fitted IDF formula was derived as:

$$I = \frac{28.5 P_{1-hr}^{Tr}}{(10 + T_d)^{0.789}} \tag{2.14}$$

in which I = rainfall intensity in inch/hour, and T_d= duration in minutes. To use Eq 2.14, the index rainfall depth is the one-hour precipitation value. For instance, to estimate the 100-yr event, the value of $P_{1\text{-hr}}$ = 2.61 inches in Table 2.5 shall be selected for the Denver area.

EXAMPLE 2.4 Estimate the 10-year 20-minute rainfall intensity and depth for the City of Denver, Colorado.

Solution: From Table 2.5, $P_{1\text{-hour}}$ = 1.61 inch for the 10-year event. Substituting the one-hr precipitation and T_d = 20 minutes into Eq 2.14 yields:

$$I = \frac{28.5 \times 1.61}{(10 + 20)^{0.789}} = 1.31 \ inch/hr$$

The corresponding rainfall depth is

$$P = I \times T_d = 1.31 \times \frac{20}{60} = 1.05 \text{ inches}$$

2.9. CLOSING

Rainfall depths are characterized by its duration and return period. For instance, the 20-minute rainfall depth represents the most intense 20-min precipitation depth. The 100-yr 20-min rainfall depth is expected to be exceeded or equaled once every 100 years. For a given return period, the precipitation depth increases with respect to the duration while the rainfall intensity decreases with respect to the duration. The relationship among rainfall intensity-duration-frequency (IDF) is best represented by a hyperbolic equation similar to Eq 2.14.

2.10. HOMEWORK

Q2.1 Figure HW 2.11 presents the complete record for 5-minute rainfall depths recorded at a gage station from 1980 to 1990.

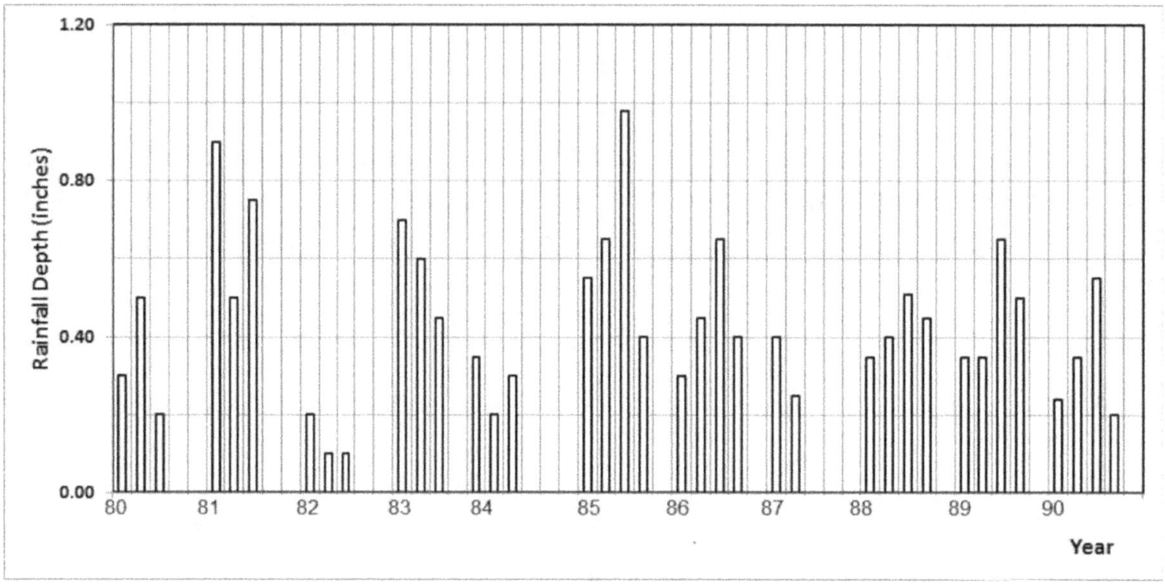

Figure HW 2.1. Complete Rainfall Data Record for AMS Study.

(a) Construct the annual maximum series

year	80	81	82	83	84	85	86	87	88	89	90
depth in inch											

(b) Rank the values in the above table.
(c) Determine the return period for the magnitude of 0.8 inch.
(d) Determine the exceedance probability for the magnitude of 0.8 inch.

Q 2.2 Explain how to derive the 100-yr 10-minute precipitation depth for a rain gage.

Q2.3 Plot the 100-year IDF formula for the Denver area, and check the discrepancy between the formula and the One-hr precipitation depths in Table 2.5.

Q2.4 The 100-year one-hr rainfall depths in the City of Boulder were obtained from the *Rainfall Atlas Volume III for the State of Colorado*. Use the empirical formulas to derive the PDF curve for the City of Boulder.

Return Yr	5-min	10-min	15-min	30-min	1-hr rainfall depth Inch
2					1.04
5					1.45
10					1.72
50					2.35
100					2.65

Q2.5 Considering that *Intensity= Precipitation/Duration*, derive IDF curve for the City of Boulder.

Return Yr	5-min	10-min	15-min	30-min	1-hr intensity in inch/hr
2					1.04
5					1.45
10					1.72
50					2.35
100					2.65

Q2.6 Derive the best-fitted equation for the rainfall IDF curve for the City of Boulder.

2.11. REFERENCES

Atlas 2 (2015), http://www.nws.noaa.gov/oh/hdsc/noaaatlas2.htm, Precipitation Frequency Estimates, NOAA

Beard, L.R., (1962). "*Statistical Methods in Hydrology*", U.S. Army Corps of Engineers, Sacramento, California.

Bulletin 17 (1982), "*Guidelines for Determining Flood Flow Frequency*", Interagency Advisory Committee on Water Data, Bulletin #17B of the Hydrology Subcommittee, OWDC, US Geological Survey, Reston, Virginia.

Chow, V.T., Maidment, David R., and Mays, Larry W., (1988) "*Applied Hydrology*", McGraw-Hill Publishing Company, New York.

Gringorten, Irving, I. (1963)."*A Plotting Rule for Extreme Probability Paper*", J. of Geophysical Research, Vol 68, N0. 3, February.

Gumbel, E. J., (1954). "*The Statistical Theory of Theory of Droughts*", Proc. Am. Soc. Civ. Eng., Vol 80, pp 439- to 439-19.

Guo, James C.Y., (1986). "*Software for Hydrologic Frequency Analysis*", the Fourth National Conference on the Application of Microcomputer to Civil Engineering, Orlando, Florida, Nov. 1986, pp 308-310.

Guo, James C.Y., Urbonas, Ben. and Stewart, Kevin. (2001). "*Rain Catch under Wind and Vegetal Effects*," ASCE J. of Hydrologic Engineering, Vol 6, No.1, in Jan.

Han, Charles T., (1977). "*Statistical Method in Hydrology*", The Iowa State University Press/Ames, 1977.

HEC SSP (2005). "*Hydrologic Statistical Package*", Hydrologic Engineering Center, US Army Corps of Engineers. Davis, California.

Peak FQ Program (2005). User's Manual for Computer Program PEAK FQ, office of surface water, USGS. http://water.usgs.gov/osw/bulletin17b/bulletin_17B.html.

SCS (1976). "*Rainfall-runoff for Small Watersheds*", TR No. 55, Soil Conservation Service (SCS) or Natural Resource Conservation Services (NRCS), U.S. Government Printing Office, Washington D.C.

UDFCD Manual (2010), "*Storm Water Drainage Design Criteria Manual*", Vol 1 and Vol 2, Urban Drainage and Flood Control District Denver, Colorado.

CHAPTER 3
STREET HYDROLOGY

Urban catchments produce flash floods because the developments in urban areas result in more impervious pavements and highly efficient drainage systems. Urban catchments are characterized as "small watershed". Although the definition of small watershed is referenced to watershed's quick response to storm event, in practice such a drainage nature is translated into the size of tributary area (AGU Committee, 2015). Among many hydrologic methods, the Rational method is the most popular storm runoff prediction procedure developed for urban catchments. As a lumped approach, the hydrologic parameters used in the Rational method are assumed to be spatially and temporally uniform across the entire catchment. In general, the assumption of hydrologic homogeneity is acceptable for a tributary area up to 100 to 150 acres.

3.1. PEAK RUNOFF PREDICTION

The peak flow rate of storm runoff generated from a small catchment is linearly related to the tributary area and rainfall intensity (Kuichling, 1889). The Rational method states:

$$Q_P = KCIA \tag{3.1}$$

in which Q_p = peak runoff in *cfs* or *cms*, I = rainfall intensity in inch/hour or in mm/hr, A = drainage area in acres or hectares, C = runoff coefficient, and K = conversion factor. The value of K is unity for the English units and 1/360 for the metric units. Today, there are many sophisticated hydrologic models developed for urban runoff predictions. The Rational method remains to be the most popular and widely used method. Although the method appears simple, care must be taken when selecting the two parameters: *runoff coefficient* and *rainfall intensity*.

Runoff coefficient represents the ratio of runoff depth to rainfall depth or a lumped parameter to represent the hydrologic losses. The value of runoff coefficient is directly related to the land uses in terms of imperviousness percentage. Table 3.1 summarizes the recommended runoff coefficients for urban areas (Guo and Urbonas 2014):

Table 3.1. Runoff Coefficients for Urban Catchments.

Land Use	Impervious Percentage	Runoff C 2-yr	Runoff C 5-yr	Runoff C 10-yr	Runoff C 100-yr
Business/Commercial	95	0.87	0.87	0.88	0.89
Business/neighborhood	70	0.6	0.65	0.70	0.8
Single-family residential	50	0.40	0.45	0.50	0.60
Multi-unit(detached)	50	0.45	0.50	0.60	0.70
Multi-unit(attached)	70	0.60	0.65	0.70	0.80
1/2 acre lot or larger	45	0.30	0.35	0.40	0.6
Apartments	70	0.65	0.70	0.70	0.80
Light Industrial	80	0.71	0.72	0.76	0.82
Heavy Industrial	90	0.80	0.80	0.85	0.9
Parks/Cemeteries	7	0.10	0.1	0.35	0.6
Playground	13	0.45	0.25	0.35	0.65
Schools	50	0.45	0.5	0.6	0.7
Railroad Yard Area	20	0.4	0.45	0.5	0.6
Paved Streets	100	0.87	0.88	0.9	0.93
Gravel Streets	40	0.15	0.25	0.35	0.65
Drive and Walks	96	0.87	0.87	0.88	0.89
Roofs	90	0.8	0.85	0.9	0.9
Lawns, sandy soil	2	0	0.01	0.05	0.2
Lawns, clay soil	2	0.05	0.1	0.2	0.4

3.2. MIXED LAND USE

An urban catchment is composed of pervious and impervious areas. The impervious percentage is defined as:

$$I_a = \frac{A_I}{A_P + A_I} = \frac{A_I}{A} \qquad (3.1)$$

Where I_a = impervious percent, A_I= impervious area in [L^2], A_P= pervious area in [L^2], and A= total area in [L^2]. When the catchment is developed into mixed land uses, the average runoff coefficient for the entire catchment is calculated by the area-weighted method as:

$$C = \frac{\sum_{i=1}^{i=n} C_i A_i}{A} \tag{3.2}$$

in which C = average runoff coefficient for watershed, A= total drainage area in [L^2] for watershed, C_i = runoff coefficient for *i-th* subarea, A_i = subarea in [L^2] and n = number of subareas.

EXAMPLE 3.1 A single-unit residential development covers an area of 400 ft by 400 ft square as shown in Figure 3.1. The 5-yr runoff coefficients for the impervious areas, porous parking surface, and lawn area are given. Determine the area-weighted runoff coefficient and imperviousness percentage.

Solution: The total area for this site is 3.67 acres which is divided into 3 different land uses as listed below:

Item	Land Use			Weighted C5	Imp% *Ia*
	Roof /Driveway	Porous Parking	Grass Area		
Area (acres)	2.30	0.83	0.55		
C5	0.90	0.60	0.30	0.74	
Imp Ia %	100.00	0.00	0.00		62.5

For this site, the average runoff coefficient is 0.74 for imperviousness of 62.5%.

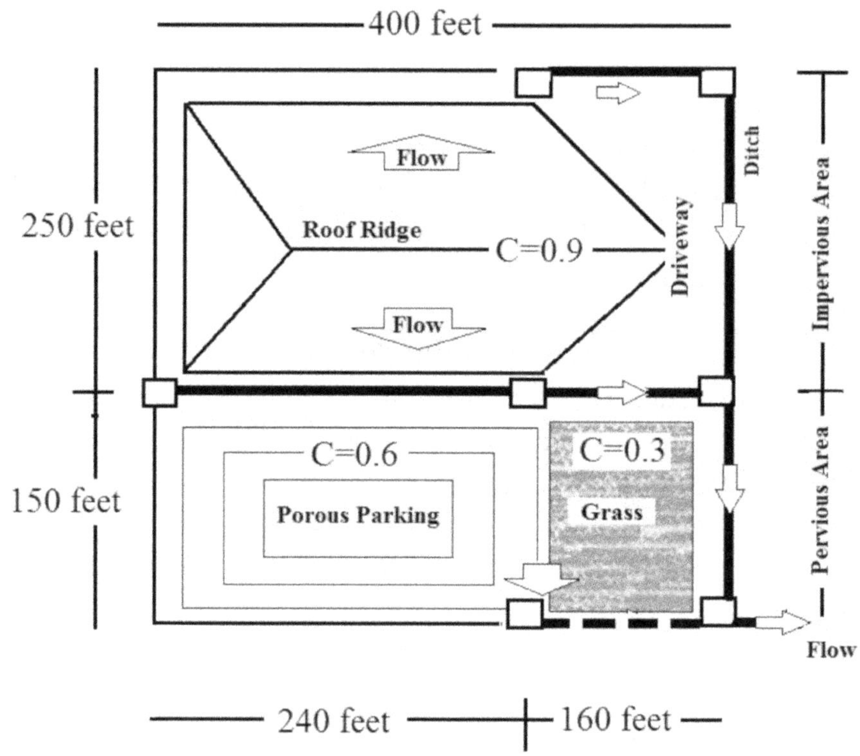

Figure 3.1. Land Uses at Residential Site.

3.3. DESIGN RAINFALL

For a small watershed, the point rainfall statistics can be directly applied to watershed runoff predictions. The Weather Bureau has published a set of rainfall statistics for the United States such as *HYDRO 35* (Frederick et. al. 1977), and *Technical Paper 40* (Hershfield 1961, 1962) that gives the rainfall frequency information for the eastern Unites States. More detailed information is also available in *Precipitation Frequency Atlas 2* for 11 western Unites States (Atlas 2, 2015). NOAA latest Atlas 14 Point Precipitation Frequency Estimates provide the latest rainfall statistics to major metropolitans in the US (Atlas 14, 2015). These publications document the empirical procedures to derive the statistical rainfall depth for a specified return period and duration. The relationship among precipitation-duration-frequency is termed *precipitation-duration-frequency* (PDF) table that can be further converted into an *intensity-duration-frequency* (IDF) table.

For convenience, the IDF formula can be derived by the best fitted equation applied to the IDF table. The IDF relationship is a decay curve, or the rainfall intensity decreases with respect to the rainfall duration. Using the City of Denver as an example, the format for IDF curve is (UDFCD Manual, 2010):

$$I = \frac{C_1 P_1}{(C_2 + T_d)^{C_3}} \quad (3.3)$$

in which P_1= index precipitation in inches, T_d= rainfall duration in minutes and C_1, C_2, and C_3 are the local rainfall constants. For instance, C_1 = 28.5, C_2= 10, and C_3=0.786 were derived for the metro Denver areas. The index precipitation reflects the rainfall return period. The City of Denver adopts the 1-hr precipitation depths as the index precipitations while the City of Las Vegas applies the 6-hr precipitation values to Eq 3.3 (CCRFCD Manual , 1999). Table 3.2 summarizes the 1-hr precipitation depths for the metro Denver area (UDFCD Manual, 2010).

Table 3.2. One-hr Rainfall Depths for Denver Areas.

P_1 inches	0.95	1.35	1.61	2.20	2.60
Return Period in years	2.00	5.00	10.00	50.00	100.00

To apply the Rational method, the design rainfall duration is assumed to be *the time of concentration* that is the travel time required for storm water to flow through the waterway from the most upstream boundary to the outlet.

$$T_d = T_c \quad (3.4)$$

Figure 3.2 illustrates three possibilities between the rainfall duration, T_d, and the time of concentration, T_c. For a long event, the entire watershed is the tributary area to the runoff generated at the outlet. For a short event, only a portion of the watershed is the tributary area. Under the linear assumption, the Rational method under a short event is revised to (Guo 2001a, 2001b):

$$Q=eKCIA \quad (3.5)$$

in which e= percentage of watershed area to be the tributary area and defined as:

$$e = \frac{T_d}{T_c} \text{ if } T_d < T_c \quad or \quad e = 1 \text{ if } T_d \geq T_c \quad (3.6)$$

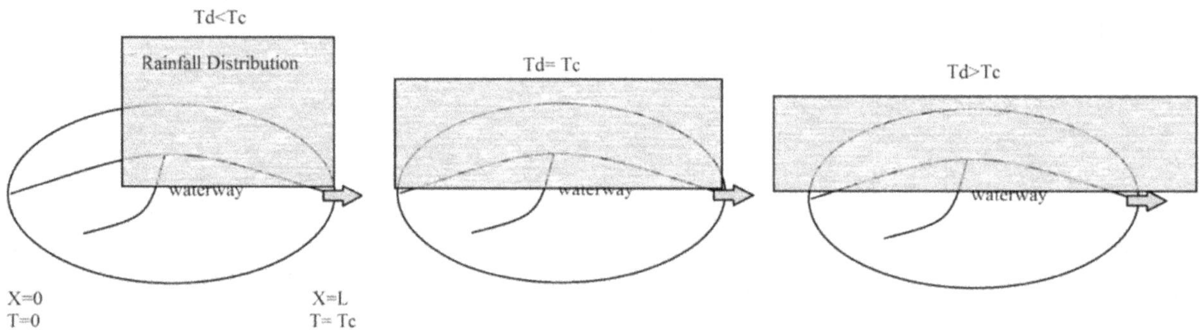

Figure 3.2. Tributary Area Proportional to Rainfall Duration.

In comparison, a long event has a less rainfall intensity and a short event has a less tributary area. Both cases underestimate the peak runoff generated from the catchment. As a result, it is concluded that the critical condition for producing the highest peak runoff is when $T_d=T_c$.

Substituting Eq 3.4 into Eq 3.3 yields the average design rainfall intensity. It is important to understand that Eq 3.3 provides the average rainfall intensity for the specified duration. In fact, the actual rainfall distribution is a continuous non-uniform distribution over the duration of T_c. Eq. 3.3 preserves the amount of rainfall, but not its distribution.

3.4. TIME OF CONCENTRATION

To calculate the time of concentration, the longest waterway length through the catchment must be considered. A waterway often begins with *overland sheet flows* for a distance less than 300 feet in an urban area or 500 feet in a rural area. Next the surface runoff will form *swale flows*. Farther downstream, a waterway may be developed with *well-defined channel* sections.

The time of concentration of the catchment is the cumulative flow times through the waterway. As illustrated in Figure 3.3, the storm water flows through a street block. The outlet is located at the street corner, Point C, to collect storm water into the downstream manhole. The longest waterway through the catchment is identified to be the length from Point A to Point C through Point B. From Point A to Point B, overland flows dominate the first 300 feet, and then the swale flow will be formed. At Point B, the street gutter will transport storm runoff to Point C.

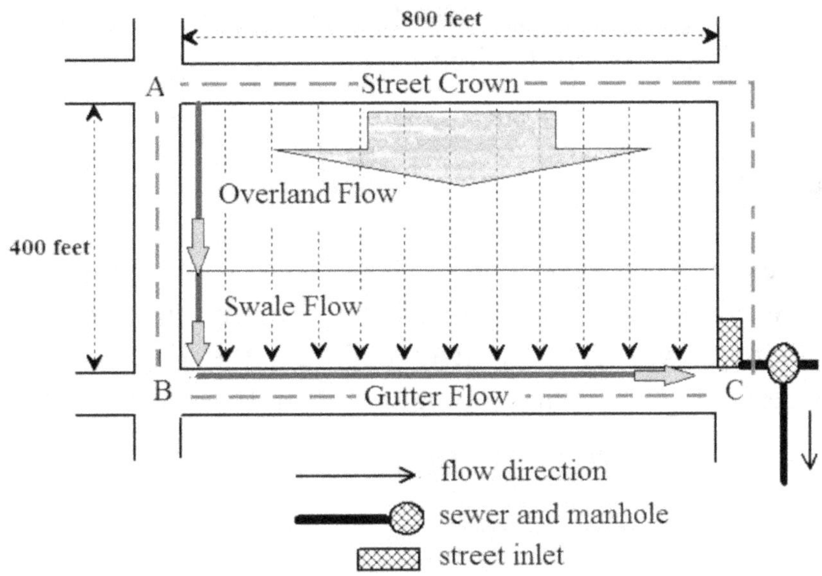

Figure 3.3. Example of Urban Catchment.

The maximum length for overland flow is recommended to be 300 feet in an urbanized area and 500 feet in a rural area. Urbanization is symbolized by curbs and gutters along streets or the 5-yr runoff coefficient≥0.2. Under a rural condition, storm water flows through natural depressions, turf strips and grass swales that significantly slow down the flow. As a result, the overland flow time in a rural area is much longer.

The catchment in Figure 3.3 is in an urban setting. The first 300 feet of Reach $\underline{AB}$ in is the overland flow length. The channel flow is in Reach $\underline{BC}$. Between the end of the overland flow and the beginning of the channel flow is the reach for swale flow. Of course, if Reach $\underline{AB}$ is shorter than 300 feet, the overland flow will directly drain into the channel flow without a swale flow.

3.4.1. Overland Flow Time

The overland flow velocity is a function of surface slope and roughness. There are many empirical formulas developed for estimating the overland flow time. For an urban setting, the *Airport formula* is recommended for calculating the overland flow time as:

$$T_o = \frac{K_{To}(1.1 - C_5)\sqrt{L_o}}{S_o^{0.33}} \tag{3.7}$$

in which T_o = overland flow time in minutes, K_{To} = 0.395 for feet-second units or 0.715 for meter-second units, L_o = overland flow length in [L], and S_o = overland slope in [L/L], and C_5 = 5-yr runoff coefficient.

3.4.2. Swale Flow Time

The Natural Resources Conservancy Services (NRCS 2015) or the former U.S. Soil Conservation Service (SCS 1976) investigated many cases from agricultural lands. The *SCS upland method* was developed to estimate the flow time through shallow swale flows. It suggests that the flow velocity through a swale be calculated as (McCuen 1982):

$$V_2 = \frac{K}{K_u}\sqrt{S_2} \tag{3.8}$$

in which V_2 = flow velocity in [L/T], S_2 = reach slope in [L/L], Ku= 1 for foot-second units or 3.28 for meter-second units, K = conveyance coefficient in [L/T] as descry bed in Table 3.3.

Table 3.3. Conveyance Coefficients for SCS Upland Method.

Linings for Conveyance	Conveyance Coefficient
Meadow (rough bushes)	2.5
Tillage (crop fields)	4.5
Lawn (grass strips)	7.0
Bare Soils	10.0
Grass Lined Swale (grass ditches)	15.0
Paved Water Way (street gutters)	20.0

A swale can be divided into several segments, depending on the variations of slope and lining material. The flow velocity through each reach is estimated by

$$T_2 = \frac{L_2}{60V_2} \tag{3.9}$$

in which T_2 = flow time in minutes, and L_2 = flow length in [L].

3.4.3. Channel Flow Time

For a well-defined channel, Manning's equation is recommended for estimating the flow velocity. However without knowing the discharge, Manning's equation is reduced to the SCS upland method as (Guo 1998):

$$V_3 = \frac{K_n}{n} R^{\frac{2}{3}} \sqrt{S_3} \approx \frac{K}{K_u} \sqrt{S_3} \qquad (3.10)$$

in which V_3 = channel flow velocity in [L/T], n = Manning's roughness, Kn= 1.486 for ft-second units or 1.0 for meter-second units, R = hydraulic radius in [L], and S_3 = channel slope in [L/L]. The value of K must be selected for the channel roughness. The flow time through a channel reach is:

$$T_3 = \frac{L_3}{60 V_3} \qquad (3.11)$$

in which T_3 = flow time in minutes, V_3 = flow velocity in [L/T], and L_3 = channel length in [L].

3.4.4. Computed Time of Concentration

Having the flow time calculated for various reaches along the longest waterway, the computed time of concentration, T_{comp}, is the sum of the flow times as:

$$T_{comp} = \sum_{j=1}^{j=N} T_j \qquad (3.12)$$

Where T_{comp}= computed time of concentration, N=number of waterways, and T_j = flow time through j-th waterway.

3.4.5. Design Time of Concentration

The time of concentration is determined based on the existing topographic condition. Without knowing the future development, the time of concentration based on the existing condition does not reflect the future hydrologic condition. For instance, waterways in a fully developed area are often equipped with drop structures for grade controls. As a result, the water flow slows down in order to avoid potential erosion on the stream bed. To be conservative, a regional formula shall be developed to represent the *future time of concentration* under the post-development condition. For instance, the Cities of Denver and Las Vegas (UDFCD 2010, CCRFCD Manual 1999) recommend the regional time of concentration be computed based on the future watershed's imperviousness ratio as (Guo and McKenzie 2014):

$$T_{Reg} = T_* + \frac{L}{60V_*} \text{ (minutes)} \tag{3.13}$$

$$V_* = \frac{K_*}{K_u}\sqrt{S_a} \tag{3.14}$$

$$T_* = 18 - 15I_a \quad (3 \leq T_* \leq 18 \textit{ minutes}, 0 \leq I_a \leq 1) \tag{3.15}$$

$$K_* = 24I_a + 12 \quad (12 \leq K_* \leq 36, 0 \leq I_a \leq 1) \tag{3.16}$$

Where T_{Reg}= regional time of concentration in minutes, L=total length of waterway in [L], including all reaches along waterway, T_*= initial overland flow time in *minutes*, V_*= post-development concentrated flow velocity in [L/T], K_*= conveyance factor in [L/T], K_u=1 for foot-second units or 3.28 for meter-second units, and S_a= average slope along waterway.

An *initial time* represents the overland flow time through the upland areas. Equation 3.15 implies that the initial time for urban overland flows is 18 minutes on a pervious surface with I_a=0, and then reduced to 3 minutes for impervious surface under I_a=1.0. The *conveyance parameter* of K_* is for shallow, concentrated flows. Equation 3.16 reveals that K_* varies from 12 to 36 *ft/sec (3.6 to 11 m/s)*, depending on watershed's imperviousness ratio. For instance, on a slope of 1%, Eq 3.16 sets the limits for the length-averaged flow velocity between 1.2 and 3.6 feet/sec with an average velocity of 2.0 ft/sec. It agrees well with the recommended *K_*=20 ft/sec* in Table 3.3 for paved surface.

In practice, the design time of concentration is the smaller one between the *computed time of concentration* representing the pre-development condition and the *regional time of concentration* representing the post-development condition as (Waugh et. al. 2002):

$$T_c = \min(T_{Comp}, T_{Reg}) \tag{3.17}$$

in which T_c= design time of concentration in minutes.

EXAMPLE 3.2 A subdivision in Figure 3.4 has an area of 400 ft by 800 ft. The land use of the subdivision is business-neighborhood or C_5= 0.65. Consider the rainfall IDF curve is:

$$I(inch/hr) = \frac{28.5 \times 1.35}{(10 + T_d)^{0.789}}$$ in which T_d= rainfall duration in minutes.

Determine the 5-year peak discharge generated from this area.

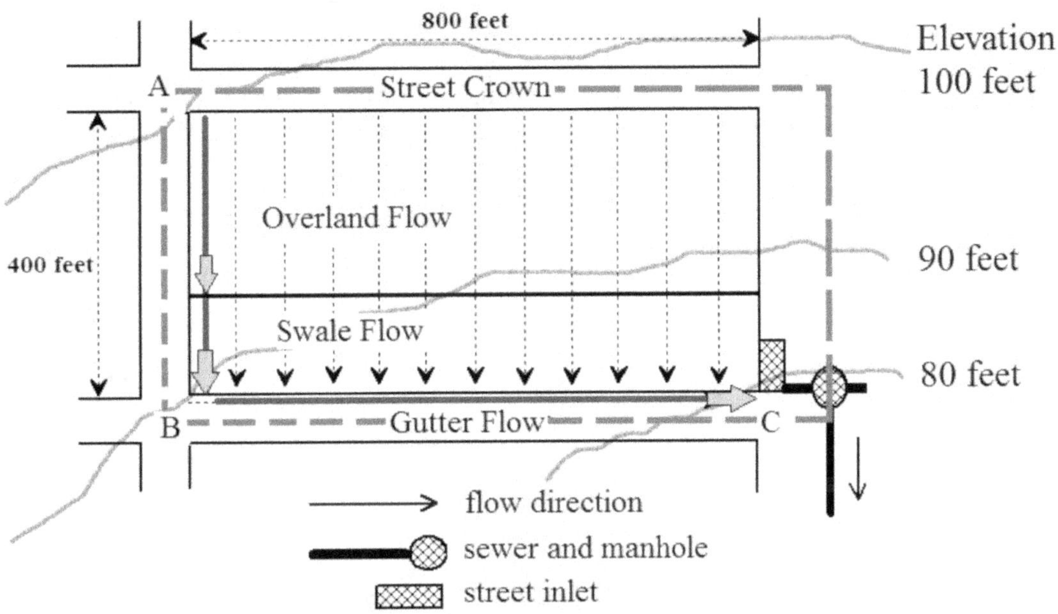

Figure 3.4. Example Catchment.

Solution

5-yr Coefficient	Overland Flow			Swale Flow			
	Slope	Length	Time	Slope	Length	SCS K	Time
	%	ft	min	%	ft	fps	min
C5	*So*	*Lo*	*To*	*S2*	*L2*	*K2*	*T2*
0.65	2.50	300.00	10.40	2.50	100.00	20.00	0.53

Gutter Flow				Existing
Slope	Length	SCS K	Time	Tc
%	ft	fps	min	min
S3	*L3*	*K3*	*T3*	*T-comp*
1.25	800.00	20.00	5.96	16.89

Drainage Area	Imp Percent	Design Runoff Coef	Total Flow Length	Average Slope	Initial Time	Convey Factor	Flow Velocity	Flow Time	Future Tc
acre		C-design	ft	%	min	fps	fps	min	min
A	*Ia*		*L*	*S*	*T**	*K**	*V**	*Tf*	*T-reg*
7.35	0.70	0.65	1200.00	1.42	7.50	28.80	3.43	5.83	13.33

Design Storm Duration	Rainfall Intensity 5-yr	Peak Flow
min	in/hr	cfs
Td	*I*	*Qp*
13.33	3.20	15.30

EXAMPLE 3.3 The peak flow in Example 3.2 was predicted under the assumption that $T_d = T_c$. Using T_c= 13.33 minutes as the basis, verify that a longer or shorter rainfall will produce a smaller peak flow.

Solution: Try two cases. Set T_d=10 minutes, e= 10/13.33=0.75 (Eq 3.6) and T_d=20 minutes with e=1.0. The peak flows are predicted as follows:

Design Storm Duration	Percent of Area	Rainfall Intensity 5-yr	Peak Flow
min		in/hr	cfs
T_d	*e*	*I*	Q_p
13.33	1.00	3.21	15.31
20.00	1.00	2.63	12.55
10.00	0.75	3.62	12.96

This example demonstrates that the conservative approach is to accept T_c as the critical T_d (Guo 2003).

3.5. MULTIPLE WATERSHEDS

The hydrologic homogeneity is one of the basic assumptions for small catchment hydrology. If the land uses or soil types in a catchment vary from one area to another, it is necessary to divide a composite catchment into sub-basins. Each sub-basin has its own outlet point. All outlets shall be connected together by waterways such as street gutters, sewers, or roadside ditches. To model the rainfall-runoff process, the physical layout of a watershed is converted into a node-link system. The drainage study of multiple subareas starts from the most upstream subarea to accumulate the flow time through the drainage network. At the n-th node on the waterway, the accumulated contributing area is

$$(A_e)_n = C_n A_n + \sum_{i=1}^{i=n-1} C_i A_i \tag{3.18}$$

The cumulative travel time through the waterway system is:

$$(T_c)_n = (T_c)_{n-1} + \frac{L_n}{60V_n} \tag{3.19}$$

in which A_e = effective contributing area in [L^2], T_c = cumulative time of concentration in minutes through the waterway system, L = waterway length in [L], V = average flow velocity in [L/T], i = i-th node upstream of the design point, and n = n-th node at the design point. When several waterways come to reach the design point, the critical rainfall duration at the design point is the longest flow time among all incoming waterways. Having known the contributing area and flow time, the peak discharge can be predicted by the Rational method.

EXAMPLE 3.4 Three basins in the City of Denver are designed to drain into a detention system in Figure 3.5. The watershed parameters for these three subareas are summarized in Table 3.4. Determine the 10-year peak discharge at point B.

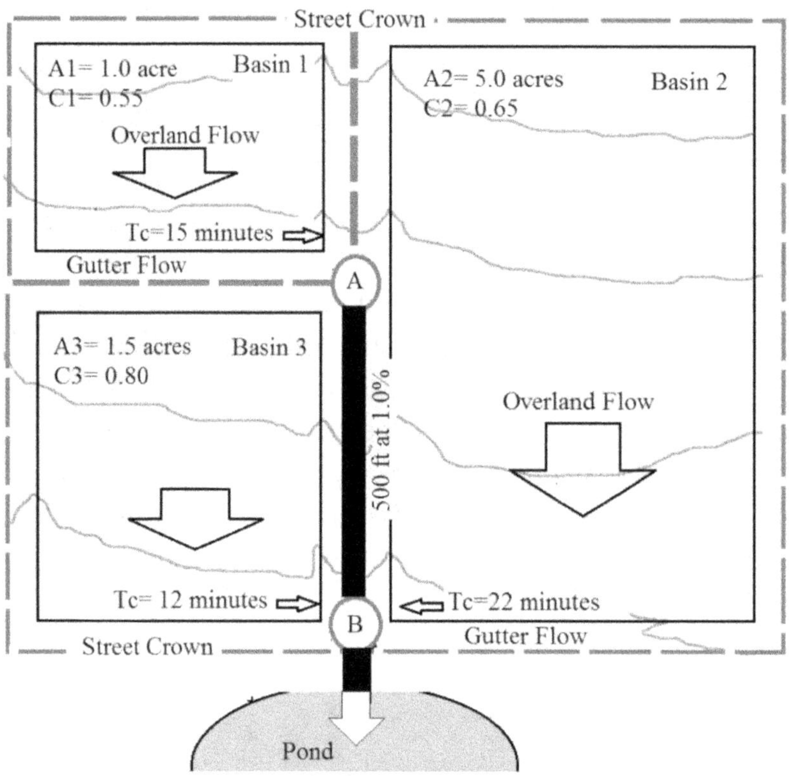

Figure 3.5. Layout for Multiple Subareas.

<u>*Solution:*</u>

Figure 3.5 indicates that there are three flow paths to reach Point B. Their flow times are:

(1) From Subarea 2: $T_C = 22$ minutes

(2) From Subarea 3: $T_C = 12$ minutes

(3) From Subarea 1: The flow time shall count for the time of concentration of Subarea 1, and the flow time from Point A to Point B through the street. According to the SCS upland method, the conveyance parameter for the paved gutter flow is 20.0 on a slope of 0.01. Aided by Eq 3.8, the flow time from Subarea 1 to Point B is the sum of the time of concentration of Subarea 1 plus the flow time through the 500-ft gutter as:

$$T_1 = 15 + \frac{500}{60 \times 20\sqrt{0.01}} = 19.17 \text{ minutes}$$

At Point B, the design rainfall duration is the longest one among three paths as:

$$T_d = \max(19.17, 22, 12) = 22 \text{ minutes.}$$

From Table 3.2, the 10-year design rainfall intensity for City of Denver is:

$$I = \frac{28.5 \times 1.61}{(10 + 22)^{0.789}} = 3.0 \text{ inch/hr}$$

According to Eq 3.18, the effective area at Point B is

$$A_e = 0.81*1.50 + 0.55*2.0 + 0.65*5.0 = 5.565 \text{ acres}$$

As a result, the 10-year peak discharge is

$$Q = I\,A_e = 16.7 \text{ cfs}$$

3.6. CLOSING

The Rational method is a linear approach. The major assumption is that the critical rainfall duration is equal to the time of concentration and the peak flow is proportional to the tributary area and average rainfall intensity. It implies that the tributary watershed must not have any significant storage effect. Secondly, the IDF formula provides a uniform intensity for a given duration. In general, the Rational method is more applicable to urban catchments up to 100 to 150 acres (Guo 2003).

3.7. HOMEWORK

Q3.1 As shown in Figure HW 3.1, the roof area produces overland flows and ditches convey the flow to the outfall. The task is to determine the peak flow from the site. Design information is given as follows:

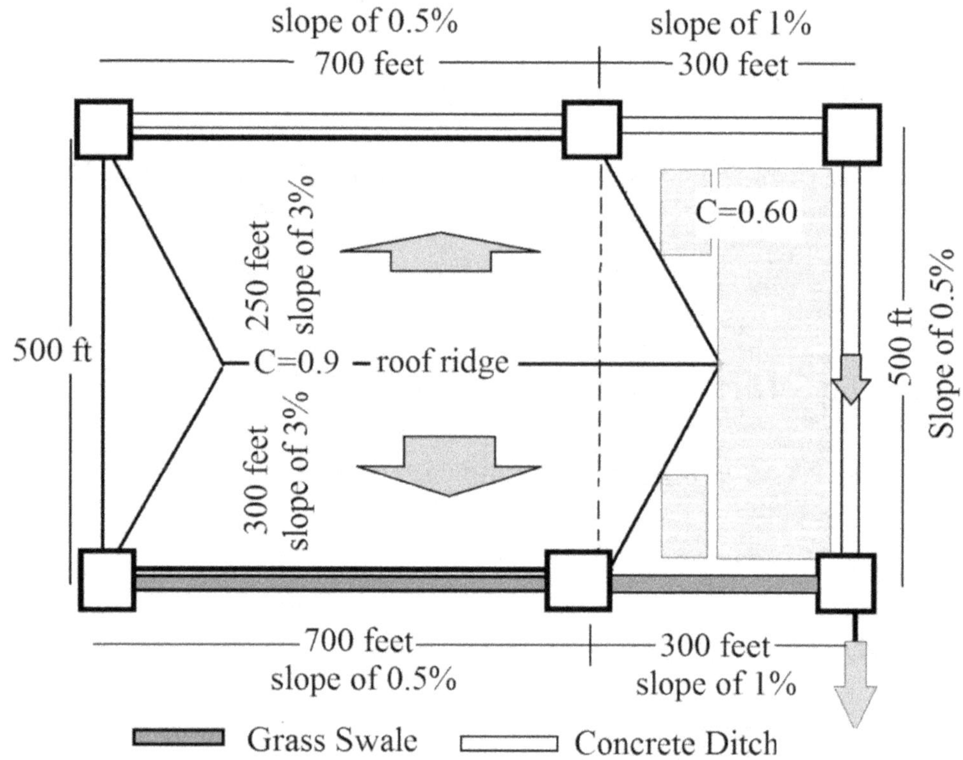

Figure HW 3.1. Residential Site for Peak Flow Prediction.

(a) Design rainfall intensity formula

$$I = \frac{38.5}{(10 + T_d)^{0.789}} \quad \text{in which } T_d = \text{duration in minutes}$$

(b) There are two flow paths through the site. The design rainfall duration for peaking runoff is set to be the longer flow time. The overland flow time shall be estimated by the airport formula, and the ditch flow velocity shall be estimated by the SCS upland method using K=15 for grass linings and 20 for concrete linings. Determine which flow path dominates the time of concentration.

(c) Use the area-weighted runoff coefficient to estimate the peak flow.

<u>Q3.2</u> A highway system in Figure HW 3.2 is built to be symmetric to the central ditch line. Strom runoff generated from both pavement surfaces drain into the central ditch lined through the grass-lined median. Use the Denver's rainfall IDF formula to estimate the 100-yr peak flow at the inlet.

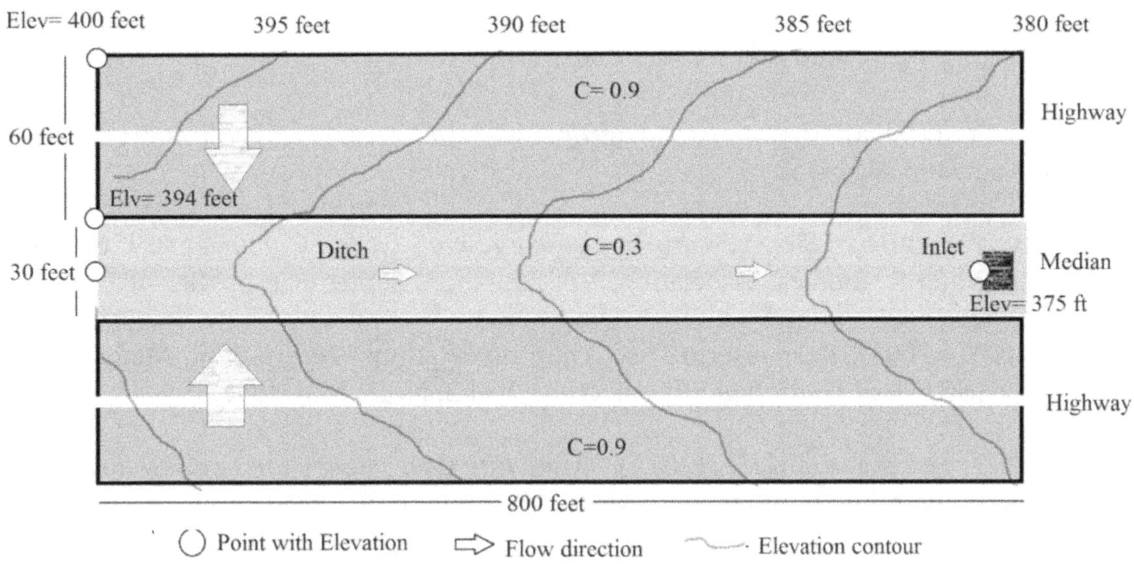

Figure HW 3.2. Highway System for Peak Flow Prediction.

3.8. REFERENCES

AGU Committee (2015). American Geophysical Union, *Committee on Runoff.* http://sites.agu.org/

Atlas 2 (2015), http://www.nws.noaa.gov/oh/hdsc/noaaatlas2.htm, Precipitation Frequency Estimates, NOAA

Atlas 14 (2015), http://hdsc.nws.noaa.gov/hdsc/pfds/pfds_map_cont.html?bkmrk=md, Precipitation Frequency Estimates, NOAA

CCRFCD Manual (1999). Clark County Regional Flood Control District, "*Hydrologic Criteria and Drainage Design Manual*", Las Vegas, Nevada.http://gustfront.ccrfcd.org/pdf_arch1/hcddm/Current%20Manual%20by%20Section/Section%20100.pdf,

Frederick, R. H., V. A. Myers, and E. P. Auciello, (1977) "*Five to 60-minute precipitation frequency for the eastern and central United States,*" NOAA Tech. Memo NWS HYDRO-35, Nat. Weather Serv., Silver Spring, Md., June.

Hershfield, D. M., (1961) "*Rainfall frequency atlas of the United States for durations from 30 minutes to 24 hours and return periods from 1 to 100 years,*" Tech. pap. 40, U.S. Dep. of Comm., Weather Bureau, Washington, D. C., May.

Hershfield, D. M., (1962) "*Extreme rainfall relationships*", J. Hydraul. Div. Am. Soc. Civ. Eng., 88(HY6), 73-92.

Guo, James C.Y. (1998). "*Overland Flow on a Pervious Surface*", International Journal of Water ,Volume 23, No 2, June, pp 91-95

Guo, James C.Y. (2001a) "*Rational Hydrograph Method",* ASCE of Hydrologic Engineering, Vol 6, No. 4, July/August, pp 352-357

Guo, James C.Y. (2001b). "*Storm Hydrographs from Small Urban Catchment*", J. of International Water, American Water Resources Association, Vol 25, No. 3, September, pp 481-487.

Guo, James C.Y. (2003). Response to Discussion on "Rational Hydrograph Method for Small Urban Catchments," ASCE J. of Hydrologic Engineering, Vol 2, No.1, May/June.

Guo, James C.Y. and MacKenzie, K (2014). "*Modeling Consistency for small to large watershed studies*", ASCE J. of Hydrologic Engineering, V 19, N8, Aug , 04014009-1 to -7.

Guo, James C.Y. and Urbonas B. (2014) " *Volume-based Runoff Coefficient*", ASCE Journal of Irrigation and Drainage Engineering, V 140, N 2, Feb, 04013013-1 to -5.

HMS (2010). "*Hydrologic Modeling Simulation Package*", Corp of Engineers, Hydrologic Engineering Center, Davis, California.

Kirpitch, Z.P., (1941). "*Time of Concentration for Small Agricultural Watersheds*," Civil Engineering, ASCE, Vol 10, No 6, June, pp 362.

Kuichling, E. (1889). "*The Relation between Rainfall and the Discharge of Sewers in Populous Districts*," Trans. ASCE, Vol 20, pp 1-56.

McCuen, R. (1982) "*A Guide to Hydrologic Analysis Using SCS Methods*", Prentice-Hall, Inc., Englewood Cliffs, New Jersey.

McCuen, R. H., Wong, S. L., and Rawls W. J. (1984). "*Estimating Urban Time of Concentration*", J. of Hydraulic Engineering, ASCE, Vol 110, No. 7, July.

NRCS(2015). SCS Upland Method,
http://www.nrcs.usda.gov/wps/portal/nrcs/site/national/home/

UDFCD Manual (2010), "*Storm Water Drainage Design Criteria Manual*", Vol 1 and Vol 2, Urban Drainage and Flood Control District Denver, Colorado.

Waugh, P.D., Jones, J.E., Urbonas, B.R., MacKenzie, K.A., and Guo, J. C.Y. (2002) "*Denver Urban Storm Drainage Criteria Manual*.", ASCE J. of Urban Drainage, Vol 112, No 56.

CHAPTER 4
STREET CONVEYANCE HYDRAULICS

Street drainage systems are vital to the urban environment. If the drainage system is not designed properly, the roadway system will be degraded to an unacceptable level of service during storm events. *Street hydraulic capacity* (SHC) includes two aspects: storm water conveyance capacity in Figure 4.1a and storm water storage capacity in Figure 4.1b. *Street Hydraulic Conveyance Capacity* (SHCC) is related to the storm water flowing on the street with a positive slope. *Street Hydraulic Storage Capacity* (SHSC) is related to the storm water storage volume in a sump area. In addition to the street slope, storm water on the street must be managed under the safety criteria that involve water flow velocity, water spread, and water depth in the gutter. Although street inlets and sewers are designed to quickly collect storm water on the street, it is important to understand that the most efficient street drainage design is to take advantage of the SHCC as the rules and regulations permit. The optimal use of SHCC can significantly reduce the number of street inlets and the sizes of storm sewers.

(a) Water Flow on Street

(b) Water Pool At Street Corner

Figure 4.1. Potential Water Hazards to Vehicles During Storm Event.

4.1. STREET HYDRAULIC CONVEYANCE CAPACITY

As indicated in Hydraulic Engineering Circulars No. 12 and No 22, entitled, "*Drainage of Highway Pavements*", the SHCC for a given street depends on street cross-sectional geometry and hydraulic roughness on the pavement surfaces (HEC 12 in 1984) (HEC 22 in 2010). In general, the water spread in Figure 4.2 for the design storm event shall not encroach into the emergency traffic lane and the gutter flow depth shall not exceed the curb height (Gallaway et al. 1979).

(a) Gutter-Full Flow	(b) Water Spread into Traffic Lane

Figure 4.2. Water Carried in Street Gutter.

4.1.1. Straight Cross Section

As illustrated in Figure 4.3, the street cross section is triangular in shape. Storm water flowing through such a triangular gutter section can be described by the revised Manning's equation as (HEC 22 in 2010):

$$Q = \frac{K}{n} S_x^{1.67} T^{2.67} \sqrt{S_o} \tag{4.1}$$

in which Q = SHCC in cfs or cms, K=0.56 for the foot-second unit system or 0.376 for the meter-second unit system, n = Manning's roughness of street surface, S_x = street transverse slope, S_o= street longitudinal slope in [L/L], and T= water spread width in [L] on the street. In general, a value of 0.016 is recommended for street Manning's roughness. The standard street transverse slope is 2%, but a slope of 1% is acceptable.

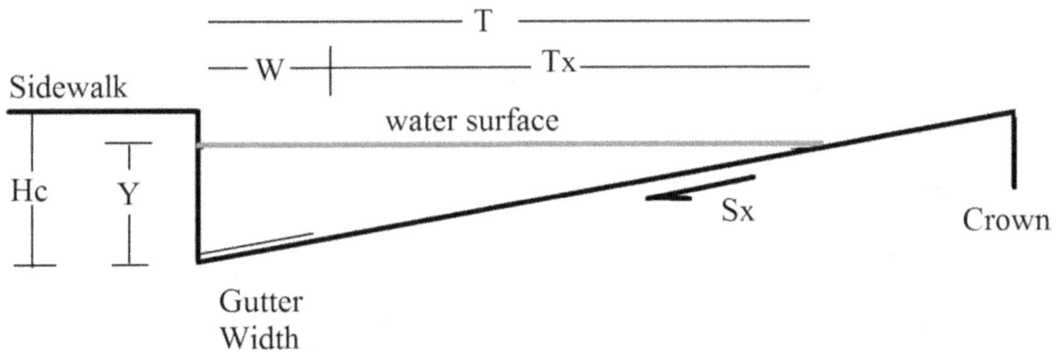

Figure 4.3. Triangular Cross Section of Gutter Flow.

The gutter flow depth, Y, at the curb is calculated as:

$$Y = TS_x \tag{4.2}$$

The cross sectional area, A, for the street flow is calculated as:

$$A = 0.5YT \tag{4.3}$$

The cross sectional average flow velocity, V, is:

$$V = \frac{Q}{A} \tag{4.4}$$

4.1.2. Composite Street Section

In practice, a standard depression of 2 inches in Figure 4.4 is often introduced at the street curb in order to increase the gutter hydraulic conveyance capacity. With an addition depression, the transverse slope across the gutter is:

$$S_w = S_x + \frac{D_s}{W} \tag{4.5}$$

in which S_W= gutter cross slope, W = gutter width such as two feet, and D_s = gutter depression such as 2 inches. The water depth at the curb is the sum of the flow depth and gutter depression as:

$$D = Y + D_s \tag{4.6}$$

in which D = water depth in gutter.

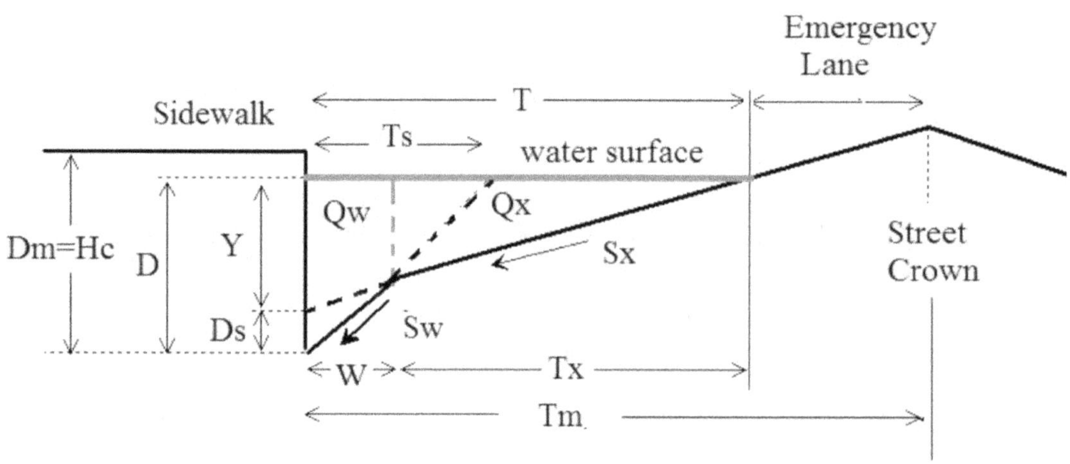

Figure 4.4. Composite Street Cross Section.

With a gutter depression in Figure 4.4, the water flow carried in a composite street-gutter cross section which is divided into *gutter flow and side flow.* The gutter flow is carried within the gutter width, W, and the side flow is carried by the water spread, T_x in [L] encroaching into the traffic lanes. For design, the water depth at a gutter must not exceed the curb height which is the maximum gutter depth, D_m in [L]. For convenience, the water spread width, T_s in [L], across the gutter width is calculated as:

$$T_s = \frac{D}{S_w} \tag{4.7}$$

The total water spread, T in [L], is the sum as:

$$T = W + T_x \tag{4.8}$$

in which T_x = water spread in [L] for side flow. Applying Eq 4.1 to both gutter and side flows, we have:

$$Q_x = \frac{K}{n} S_x^{1.67} T_x^{2.67} \sqrt{S_o} \tag{4.9}$$

$$Q_w = \frac{K}{n} S_w^{1.67} [T_s^{2.67} - (T_s - W)^{2.67}] \sqrt{S_o} \tag{4.10}$$

in which Q_X = side flow in $[L^3/T]$, Q_W = gutter flow in $[L^3/T]$, W= gutter width in [L], and T_S = water spread in [L] defined in Figure 4.3. The total flow, Q in $[L^3/T]$, on the street is

$$Q = Q_x + Q_w \tag{4.11}$$

The flow cross sectional area in $[L^2/T]$ for a composite street is

$$A = 0.5YT + 0.5WD_s \tag{4.12}$$

Aided by Eq's 4.11 and 4.12, the flow velocity on a composite street is calculated by Eq 4.4.

<u>EXAMPLE 4.1</u> Referring to Figure 3.4 in Chapter 3, the street BC has a composite section as illustrated in Figure 4.5. Street BC has T_m=15 feet, n =0.016, W=2 feet, D_s=2 inches, S_o=1.50 %, and S_x=2%. The maximum gutter depth, D_m, for this section is equal to the cub height, H_c, of 6 inches. During a minor event, the street crown must be free from water or the maximum water spread is 15 feet for this case. Determine the street-full capacity for this street.

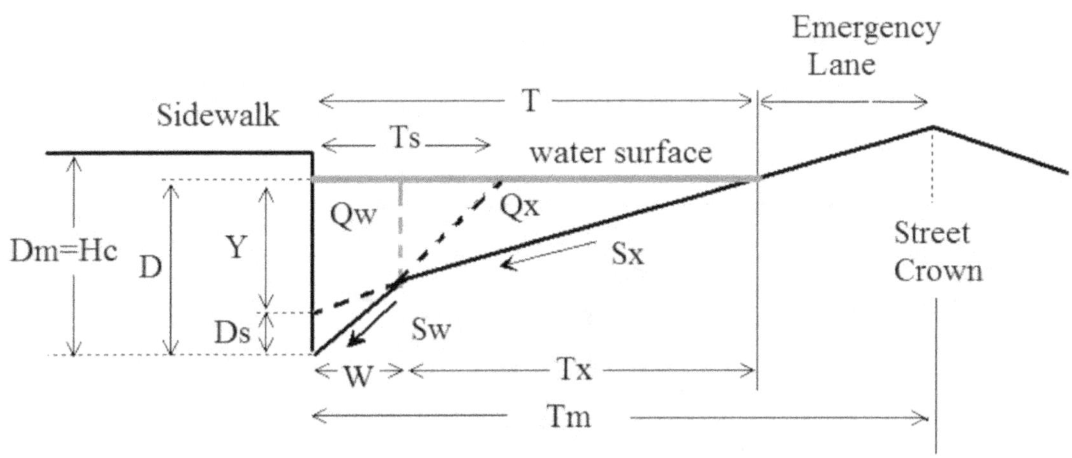

Figure 4.5. Street-full Flow Capacity.

<u>Solution</u>: Referring to Figure 4.5, the specified water spread, T =15 feet, for this case. The street flow is calculated as:

$S_w = S_x + D_s/(12W) = 0.02 + (2/12)/2 = 0.103\ ft/ft,$

$$T = \frac{(D_m - D_s)/12}{S_x} = \frac{(Y-2)/12}{0.02} = 15\ ft\ , So\ Y=0.3\ ft\ and\ D=0.47\ ft < D_m\ (Ok)$$

$T_x = T\text{-}W = 15.0 - 2.0 = 13.0\ ft,$

$T_s = 6.0/(12*0.103) = 4.52\ feet$

$$Q_w = \frac{0.56}{0.016}(0.103)^{1.67}[4.52^{2.67} - (4.52-2.0)^{2.67}]\sqrt{0.015} = 5.88\ cfs$$

$$Q_x = \frac{0.56}{0.016} 0.02^{1.67} 13.0^{2.67} \sqrt{0.015} = 4.28\ cfs$$

$Q = Q_x + Q_w = 10.16\ cfs$

EXAMPLE 4.2 Referring to Figure 4.6, the composite street section has T_m=15 feet, n =0.016, W=2 feet, D_s=2 inches, S_o=1.50 %, and S_x=2%. The maximum gutter depth, D_m, for this section is equal to the cub height, H_c, of 6 inches. During a minor event, the street crown can be under water for this case. Determine the gutter full capacity for this street.

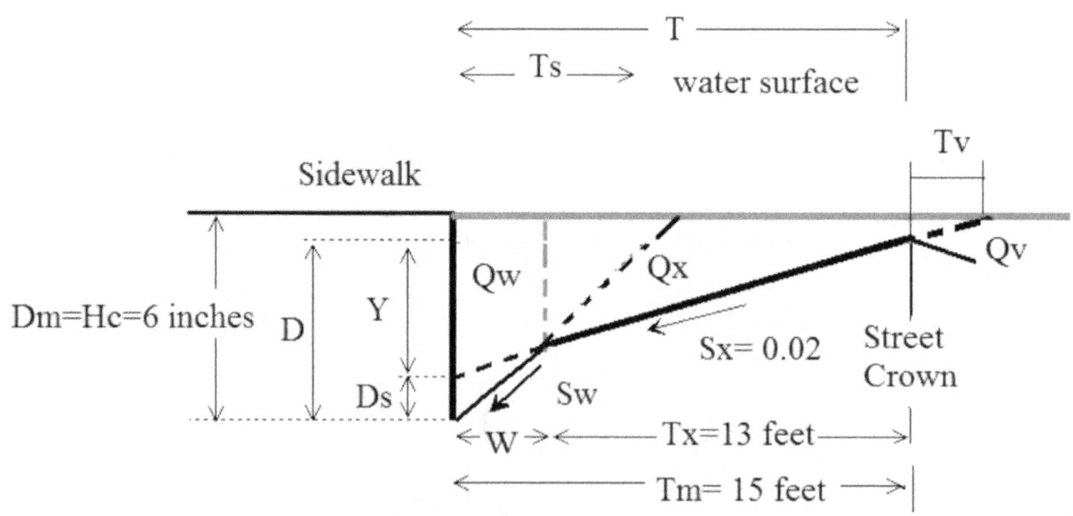

Figure 4.6. Gutter Full Flow Capacity.

Solution: Let D =D_m= H_c= 6.0 inches. For this case, the gutter full condition results in a water spread>Tm (15 ft). It means that the street crown is under water and a *virtual flow*, Qv, as illustrated in Figure 4.6, needs to be subtracted from the side flow.

$S_w = S_x + D_s/(12W) = 0.02 + (2/12)/2 = 0.103$ *ft/ft,*

The gutter full water spread width is calculated as:

$$T = \frac{(D_m - D_s)/12}{S_x} = \frac{(6-2)/12}{0.02} = 16.67\ ft, \quad > 15\ feet$$

$T_c = 16.67 - 15.0 = 1.67$ *feet.* So, a virtual flow exists.

$T_x = T - W = 14.67$ *ft,*

$T_s = 6.0/(12*0.103) = 4.85$ *feet*

$$Q_W = \frac{0.56}{0.016}(0.103)^{1.67}[4.85^{2.67} - (4.85-2.0)^{2.67}]\sqrt{0.015} = 4.95\ cfs\ (Gutter\ Flow)$$

$$Q_X = \frac{0.56}{0.016}0.02^{1.67}14.67^{2.67}\sqrt{0.015} = 8.11\ cfs\ (Side\ Flow)$$

$$Q_V = \frac{0.56}{0.016}0.02^{1.67}1.67^{2.67}\sqrt{0.015} = 0.03\ cfs\quad (Virtual\ Flow)$$

$Q = Q_X + Q_W - Q_V = 13.03$ *cfs (subtracting the virtual flow from the side flow.)*

4.2. TRAFFIC SAFETY WITH RUNOFF ON STREET

The hydroplaning effect is associated with the spinning acceleration of a vehicle on a thin layer of water (Huebner, Reed and Henry 1986). The hyperbolic relationship between the flow velocity, V, and the water depth, D, sets a safety limit for the VD product. With D_s =0, the VD product can be derived from Eq 4.1 as:

$$VD = \frac{2K}{n}(TS_x)^{1.67}\sqrt{S_0} \tag{4.13}$$

$$Q = VA \tag{4.14}$$

Aided with Eq 4.3, Eq 4.14 is converted to:

$$Q = V(\frac{1}{2}DT) = \frac{1}{2}(VD)T \tag{4.15}$$

Equation 4.15 describes the relationship between SHCC and VD product. Aided with Eq 4.15, the unit-width flow, q in [L^2/T], across the water spread is:

$$q = \frac{Q}{T} = \frac{VD}{2} \tag{4.16}$$

The momentum impulse of a gutter flow consists of both static and dynamic force components. In comparison, the static force in a gutter flow is negligible because of the shallow depth. Aided by Eq's 4.1 and Eq 4.15, the dynamic force, M, associated with a gutter flow is

$$M = \rho QV = \frac{\rho}{2S_x}(VD)^2 \tag{4.17}$$

in which M = dynamic force, and ρ = density of water. Equations 4.16 and 4.17 indicate that both SHCC and dynamic force in a gutter flow are proportional to its VD product. A reduction on the VD product can directly impose a limit on both the unit-width capacity and momentum impulse in a gutter flow. Consider that the VD product of a gutter flow shall not exceed a limit, L, defined by safety as:

$$VD \leq L \tag{4.18}$$

in which L= permissible VD product. For instance, L = 6.0 cfs/ft recommended by Clark County in Las Vegas, Nevada for a 10-year event. According to Eq 4.15, Clark County imposes a limit of 3.0 cfs per foot to the gutter flow, and the momentum force carried by the gutter flow section becomes no more than 1,728.00 pounds when S_X = 0.02 ft/ft, and ρ = 1.92 slug/cubic ft. Substituting Eq 4.13 into Eq 4.18 yields the permissible water spread, T_L, for the specified L as (Guo 2000):

$$T_L \leq \frac{1}{S_x}\left(\frac{nL}{2K\sqrt{S_0}}\right)^{0.6} \tag{4.19}$$

Aided with Eq 4.15 and Eq 4.18, the allowable street conveyance capacity (ASHCC) is equal to a flow reduction as:

$$Q_L = \frac{1}{2} L T_L \tag{4.21}$$

in which Q_L= allowable street hydraulic capacity in $[L^3/T]$.

4.3. DISCHARGE REDUCTION METHOD

In parallel to the approach of permissible VD product, the discharge reduction method is also developed and recommended for street drainage designs. For instance, the City and County of Denver converts the limiting VD product into a set of reduction factors. The allowable SHCC is equal to the gutter-full capacity multiplied by a reduction factor that is defined as (Guo 1997):

$$R = \frac{Q_L}{Q_{GF}} \tag{4.22}$$

in which Q_{GF} =gutter full SHCC and R = reduction factor. Substituting Eq's 4.19, 4.21 into Eq 4.22 yields

$$R = \frac{\frac{1}{2} L T_L}{V\left(\frac{1}{2} D_m T\right)} = \frac{1}{(TS_x)^{2.67}} \left[\frac{nL}{2K\sqrt{S_0}}\right]^{1.60} \quad \text{for } 0<R<1 \tag{4.23}$$

As shown in Eq 4.23, the flow reduction factor is reversely proportional to the street longitudinal slope for specified L and T. In the Metro Denver, Eq 4.23 was used to produce a regional safety chart for flow reductions on steep streets. Considering T=12.5 feet and L=1.0 cfs/ft for the minor event and T=20.5 feet and L=2.0 cfs/ft for the major event, Figure 4.7 was produced for street longitudinal slopes ranging from 0.5 to 10% (Guo 1999). For a steep street, Eq 4.23 results in R<1.0, indicating that the design flow on the street shall be reduced. For streets with a mild slope, Eq 4.23 may result in R>1.0, indicating that the street gutter can carry more than the design discharge without exceeding the permissible VD product.

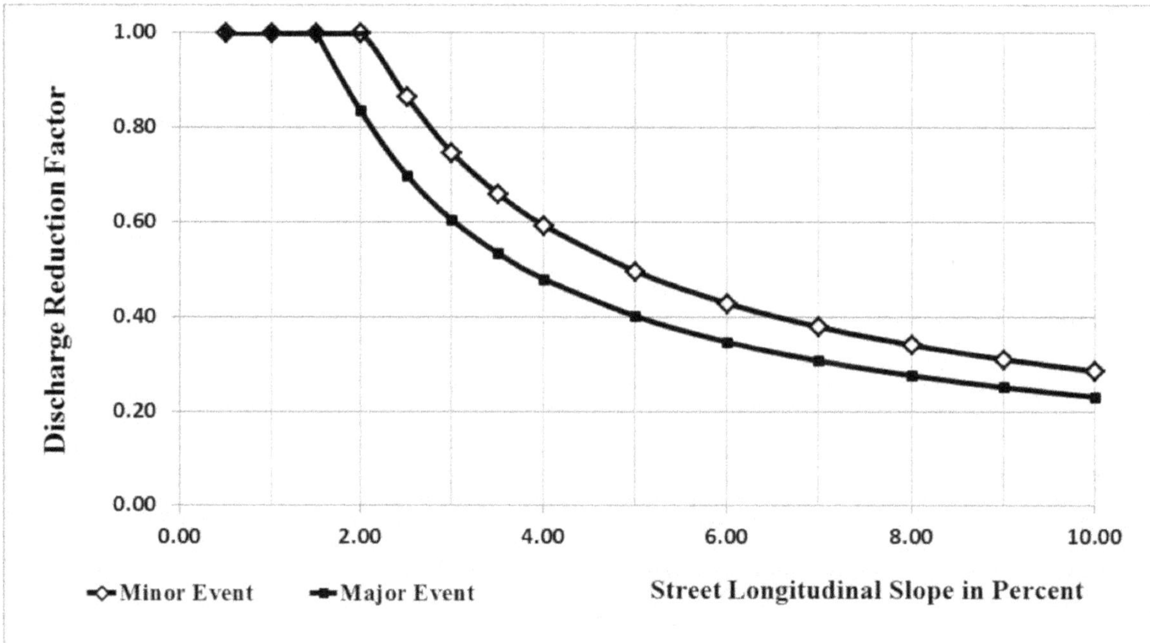

Figure 4.7. Reduction Factors For Street Flow Capacity.

4.4. ALLOWABLE STREET HYDRAULIC CONVEYANCE CAPACITY

For a given street cross section, the street full flow capacity, Q_{SF}, is firstly determined by the specified water spread. Secondly, the gutter full capacity, Q_{GF}, shall be calculated based on the curb height, and the flow reduction factor is determined from Figure 4.7. The allowable SHCC is determined as (UDFCD Manual 2010):

$$Q_a = \min(Q_{SF}, RQ_{GF}) \tag{4.24}$$

In practice, it is critical to know the allowable SHCC at the street site because wherever the water flow on the street exceeds the allowable defined in Eq 4.24, an inlet or a culvert must be installed to remove the storm runoff away from the street.

EXAMPLE 4.3 Referring to Figure 3.4, Street BC was evaluated for both the street full hydraulic conveyance capacity in EXAMPLES 4.1 with T_m= 15 feet and the gutter full hydraulic conveyance capacity in EXAMPLE 4.2 with D_m= 6 inches. Determine the allowable flow capacity for this street.

Q_{SF} =10.16 cfs from EXAMPLE 4.1

Q_{GF}= 13.03 cfs from EXAMPLE 4.2

CHAPTER 5
STREET STORAGE CAPACITY

Along a straight street section, the hydraulic conveyance capacity is determined by Manning's formula (Guo 2000a). At a street intersection as shown in Figure 5.1, the two street crowns form a depressed pool at the street corner. Under the concern of traffic safety, it is preferable that the storm water accumulated at the street intersection is not allowed to flow across the streets. In other words, the headwater provided by the curb height shall be sufficient to drain storm runoff into the sump inlets at the street intersection. The water spread at a street intersection is solely dictated by the geometry at the street intersection rather than the conveyance capacity. Therefore, *a volume-based method* shall be derived to predict the water spread at a sump like a street intersection (Guo 2000b). Drainage design at a sump such as a street intersection must be determined with the volume balance between (1) *storm water detention volume*, and (2) *storage capacity* around the sump inlet.

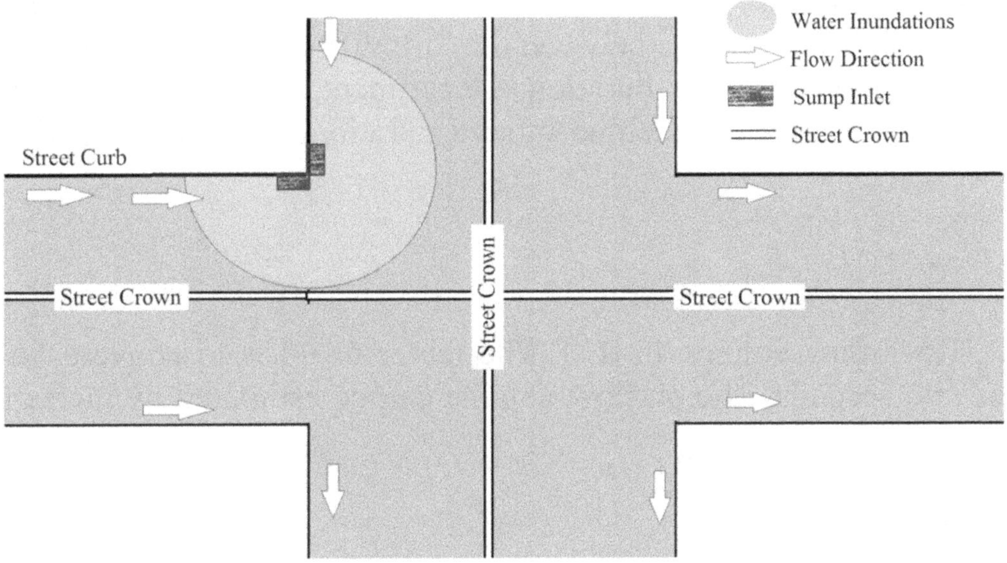

Figure 5.1. Water Inundations at Sump at Street Intersection.

5.1. STORM WATER DETENTION VOLUME

In an urban area, the distance between two adjacent street inlets is approximately 300 to 400 feet. The catchment area for a sump inlet is approximately one to three acres. To predict the peak runoff from a small urban catchment, the Rational method states:

$$Q_d = KCI_d A \tag{5.1}$$

The rainfall intensity in Eq 5.1 can be described as:

$$I_d = \frac{a}{(T_d + b)^m} \tag{5.2}$$

n which Q_d= peak runoff rate in cfs (cms), K = unit conversion factor, equal to 1 for English units, and 1/360 for SI units, C = runoff coefficient, A= watershed area in acres (hectares), I_d = rainfall intensity in inch/hr or mm/hr, T_d = rainfall duration in minutes, and a, b, and m= constants on the Intensity- Duration- Frequency (IDF) formula. Under an event, storm runoff flows toward the sump areas where the accumulated storm water will drain into the sewers. The basic concept in the volume-based method is to find the maximum volume difference between the inflow and outflow volumes under a series of storm events with different duration (Department of the Army and the Air Force in 1977, Guo in 1999). For specified rainfall duration, the inflow runoff volume is determined as:

$$V_i = 60KCAI_d T_d \tag{5.3}$$

Where V_i = inflow volume in [L^3]. The factor of 60 is to convert the duration, T_d, in minutes into seconds. The outflow volume can be estimated by the sump inlet capacity as

$$V_o = 60QT_d \tag{5.4}$$

in which V_o = outflow volume in [L^3], and Q = sump inlet capacity in [L^3/T] designed to pass a 2- or 5-year peak discharge. Aided by Eq's 5.1, 5.2, 5.3 and 5.4, the storm water detention volume, V_d, is obtained as

$$V_d = 60[KCA\frac{a}{(T_d + b)^m}T_d - QT_d] \tag{5.5}$$

To maximize the storage volume, the first derivative of Eq 5.5 with respect to rainfall duration is set to be zero. It yields:

$$\frac{dV_d}{dT_d} = [\frac{-mT_d}{(T_d+b)^{m+1}} + \frac{1}{(T_d+b)^m} - \frac{Q}{KaCA}]_{T_d=T_m} = 0 \tag{5.6}$$

Solution of Eq 5.6 is:

$$T_m = \frac{1}{m}[(T_m+b) - \frac{Q}{KaCA}(T_m+b)^{m+1}] \tag{5.7}$$

in which T_m = design rainfall duration in minutes. By trial and error, Eq 5.7 is solved for T_m. Substituting $T_d = T_m$ into Eq 5.5, the design storm water detention storage volume can be maximized. Of course, such a maximal solution can also be achieved by a finite difference approach. For instance, use an increment of one minute on rainfall duration to test Eq 5.5 from T_d = 5 through 60 minutes until the maximal value is achieved.

5.2. STREET DEPRESSION STORAGE CAPACITY

As illustrated in Figure 5.2, the storage volume around a sump inlet consists of two portions: volumes below and above the curb height. The depression storage volume at a street corner is similar to an inverted conic volume with the street transverse slope as the conic side slope. The conic volume between two adjacent depths is calculated as (HEC-1 in 2005):

$$V_h = \frac{1}{3}h_1A_1 + \frac{(h-h_1)}{3}(A + A_1 + \sqrt{AA_1}) \tag{5.8}$$

in which h = water depth in [L], A = water surface area in [L^2], V_h= conic volume in [L^3] at depth, h. The subscript 1 represents the variables at depth, h_1. When h_1=0, i.e. on the ground, its area, A_1=0. Equation 5.8 is reduced to:

$$V_h = \frac{1}{3}hA \tag{5.9}$$

As illustrated in Figure 5.2, the storage volume around a curb is sum of two portions: *(a) volume below the curb height, H_c, and (b) the volume above the curb height.*

(a) Volume for $h<H_c$

As illustrated in Figure 5.3, when the depth, h, is less than the curb height, the water surface area can be approximated as a fraction of a circle, depending on the configuration of the street section. For instance: k= 1/4 at a 90-degree depressed street section, k = 1/2 in Figure 5.4a for a straight depressed street section, k= ¾ in Figure 5.4b for a 270-degree intersection area of two sloping streets, and k =1 under the condition of no curb. With a specified k, the water surface area at a depth, $h < H_c$ is:

$$A_h = k\pi R_h^{\ 2} \qquad \text{for } h<H_c \qquad (5.10)$$

in which k = a faction of a circle, and R_h = radius in [L] of the circular water area at a sump.

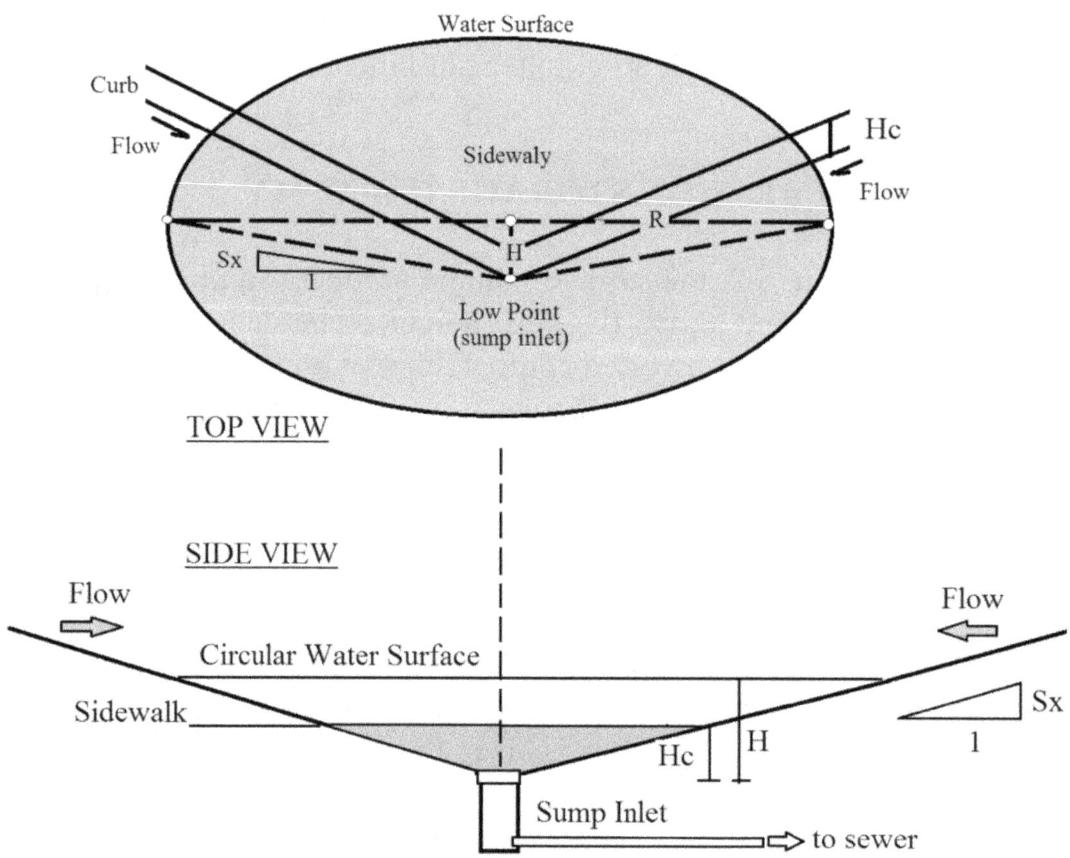

Figure 5.2. Illustration of Street Depression Storage Volume.

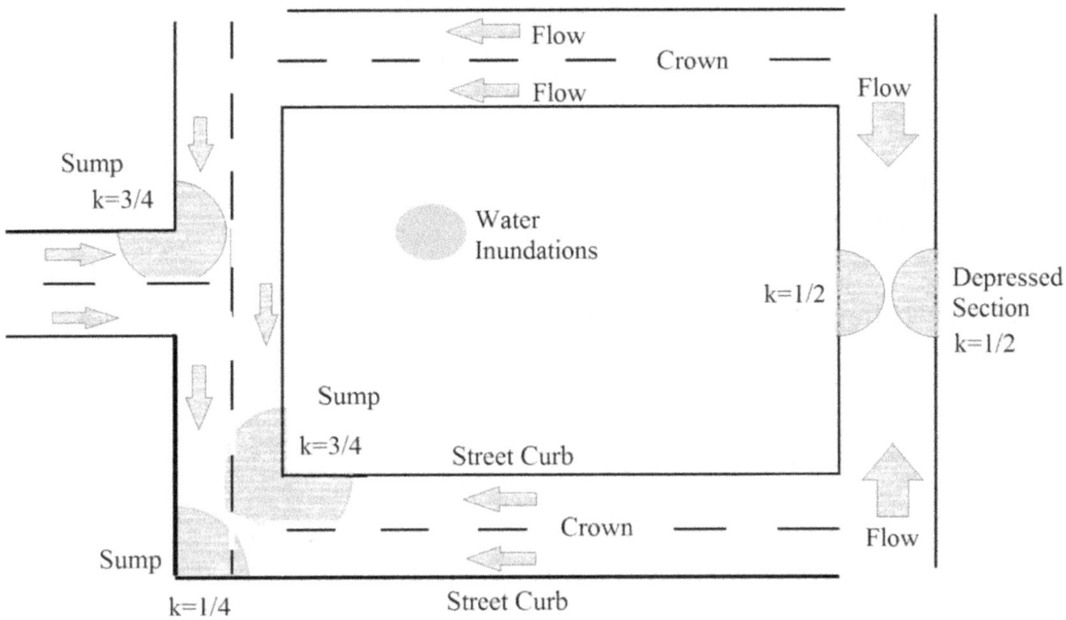

Figure 5.3. Storage Parameters at Sumps and Depressions.

(a) Ponding water with K=1/2	(b) K=3/4 for the pool water

Figure 5.4. Storage Parameters at Sumps and Depressions.

The radius of the water surface is approximated by the water depth, h, and the street transverse slope, S_X, as:

$$R_h = \frac{h}{S_x} \quad (5.11)$$

Substituting Eq's 5.10 and 5.11 into Eq's 5.9 yields

$$V_{h<Hc} = \frac{k}{3}\frac{\pi h^3}{S_x^2} \qquad \text{for } h<=H_c \qquad (5.12)$$

(b) Volume for h>H$_c$

Above the curb height, the additional volume between h and H_c is estimated by Eq 5.8 as:

$$V_{h>Hc} = \frac{\pi(h-H_c)}{3S_x^2}(H_c^2 + h^2 + H_c h) \qquad (5.13)$$

Aided by Eq 5.12, the total storage volume for a depth, h, above the curb height is:

$$V_h = \frac{k}{3}\frac{\pi H_c^3}{S_x^2} + \frac{\pi(h-H_c)}{3S_x^2}(H_c^2 + h^2 + H_c h) \qquad \text{for } h>H_c \qquad (5.14)$$

Where $V_{h<Hc}$ = water volume in [L^3] below H_c, $V_{h>Hc}$ = water volume in [L^3] above H_c, and V_h = water volume at depth, h in [L].

After knowing the detention volume by Eq's 5.5 and 5.7, the corresponding water depth is predicted by Eq 5.12 or 5.14, depending on the water depth relative to the curb height. The corresponding spread, R, can be predicted by Eq 5.11. Considering the standard transverse slope of 0.02, Figure 5.5 presents the plot of Eq's 5.12 and 5.14 with H_c= 6 inches (Guo 2000b).

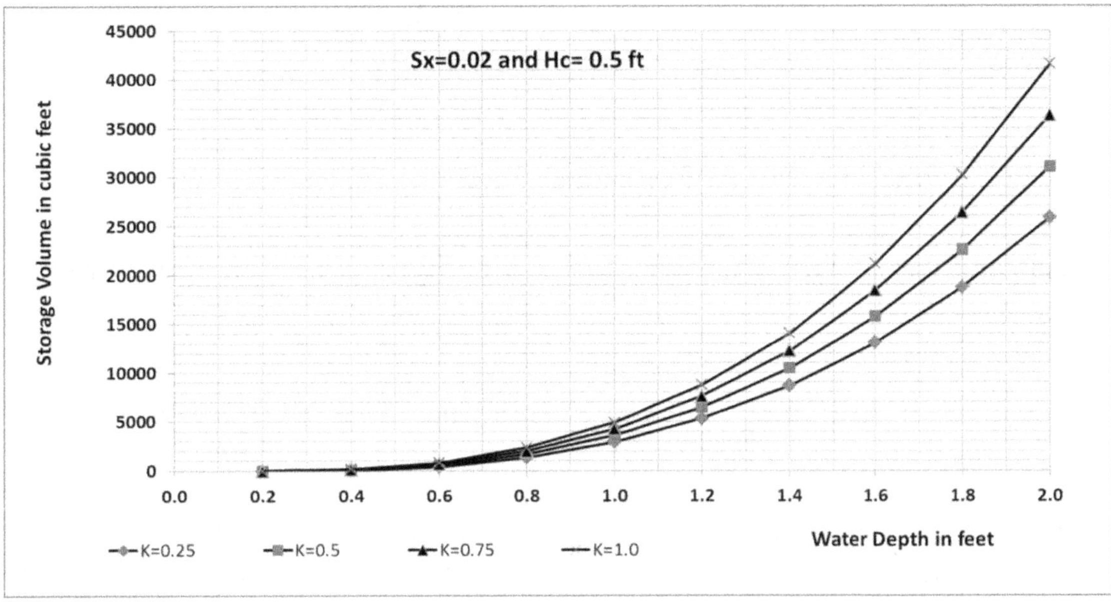

Figure 5.5. Street Storage Capacity with $S_x = 0.02$ and $H_c = 0.5$ ft.

EXAMPLE 5.1 As shown in Figure 5.6, the depressed section on South Speer Avenue in the City of Denver, Colorado has a width of 100 feet and length of 1000 feet. At the low point, a sump trench collects storm runoff from both directions. The capacity of this trench inlet is 6.5 cfs. For safety, the maximum water depth at the sump is no more than 12 inches. Evaluate if the water ponding at the low point under the 100-yr event satisfy the design criteria. The design rainfall intensity formula is given as:

$$I = \frac{74.1}{(10+T_d)^{0.789}}$$ in which T_d = duration in minutes.

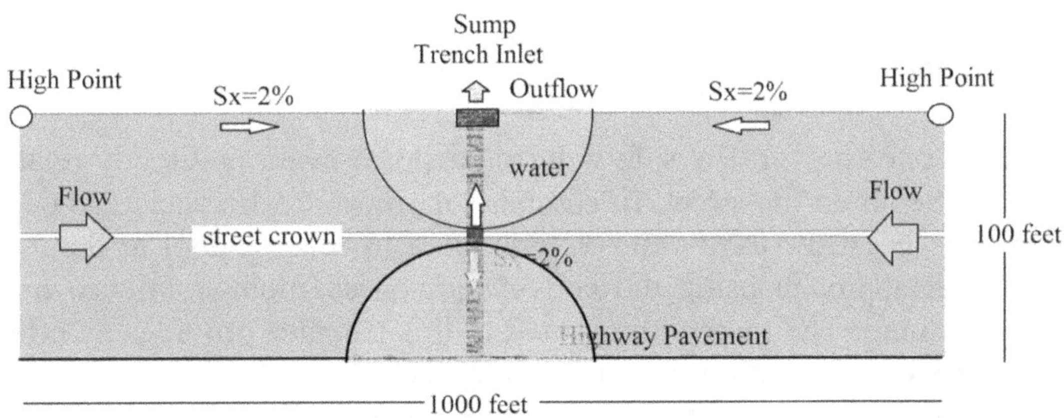

Figure 5.6. Depressed Street Section along South Speer Avenue, Denver, Colorado.

Solution: Q=6.5 cfs, K=1.0 for English units, C=0.90, and A= 2.3 acres. Equation 5.7 becomes:

$$T_m = \frac{1}{0.789}[(T_m+10) - \frac{6.5}{74.1\times1.0\times0.90\times2.3}(T_m+10)^{0.789+1}]$$

By trial and error, T_m= 14.6 minutes. Aided by Eq 5.5, the storm water detention volume is calculated as:

$$V_d = 1.0\times0.90\times2.3\times\frac{74.1}{(14.6+1.0)^{0.789}}\times14.6\times60 - 6.5\times14.6\times60 = 5042\ ft^3$$

The storm water volume of 5042 ft^3 will be accumulated at the sump inlet. Next, we need to estimate the available storage volume at the site. Aided with Eq 5.14, the water depth at the pool is:

$$5042 = 2[\frac{0.50}{3}\frac{\pi\times0.5^3}{0.02^2} + \frac{\pi(h-0.5)}{3\times0.02^2}(0.5^2+h^2+0.5\times h)]$$ So, h= 0.9 foot<1.0 ft.

The water depth for this case is 0.9 foot and the corresponding radius for the water pool is:

$$R_h = \frac{0.9}{0.02} = 45 < 50\ feet.(Ok!)$$

For comparison, Eq 5.14 was also analyzed for various side slopes. It was predicted to have a ponding depth of 1.07 feet for S_x=0.03 or 0.6 foot for S_x=0.01.

5.3. CLOSING

The major concern of traffic safety during a storm event is directly related to the water spread and depth on the street. In comparison, more traffic accidents occur at the street intersections because storm water is entrained to form a pool at the sump inlet. The conventional approach using the conveyance-based method fails to predict the water spread at a sump. The method presented in this Chapter provides an effective means to evaluate the water ponding and maximum accumulated depth at a street interception. Especially a depressed street section must be evaluated the water ponding situation under a major event. The method indicates that the flatter the surface slope, the shallower the accumulated water depth.

5.4. HOMEWORK

Q5.1 A curb-opening inlet as shown in Figure HW 5.1 is located at a street corner. The tributary area to this inlet is 2.58 acres with a runoff coefficient of 0.55. This inlet is protected from urban debris with blocks around the curb opening. The seepage flow into the inlet is 2.5 cfs before the water flow overtops the blocks. Determine the height of the blocks for the 5-yr Denver's storm event.

Solution

For this case, the surface area of the pooled water is K=3/4 of a circle. The rainfall IDF is prescribed with a= 38.5, b=10, and m=0.789 for the 5-yr event. The critical rainfall duration is solved for T_d= 9.07 minutes and the detention volume is 817.54 ft^3. Considering the transverse slope 0f 0.02, the block shall be designed with a height of 0.75 foot which results in a water spread of 37.5 feet.

(a) Street Inlet with Blocks

K=1 for ft or K=1/360 for meter	K=	0.75	
Tributray Area	A=	2.58	acre
Runoff Coefficient	C=	0.550	
Outflow from Sump Inlet	Q=	2.50	cfs
Rainfall Information	a=	38.50	
	b=	10.00	
$I_d = \frac{a}{(T_d + b)^m}$	m=	0.789	
Guess Duration Tm	Tm=	9.07	minutes
Check if Tm is acceptable	Balance=	0.000	
Design Rainfall Intensity Td=Tm	I=	3.76	inch/hr
Stormwater Detention Volume	V-design=	817.54	cubic ft
Storage Volume in Sump			
Street Transverse Slope	Sx=	0.02	ft/ft
Curb Height	Hc=	1	feet
Incremental Depth	Dy=	0.25	foot

Depth h (ft)	STORAGE Vh		VOLUME cubic ft
K=	0.25	0.50	0.75
0.00			0.00
0.25			30.68
0.50			245.44
0.75			815.17

(b) Ponded Water Depths and Volumes

Figure HW 5.1. Design of Sediment Bay at Street Corner.

Q5.2 A highway runs through an alluvial fan in Figure HW 5.2 that has an area of 4 acres with a runoff coefficient of 0.6. The rainfall IDF formula at this site is:

$$I = \frac{45.5}{(6.5+T_d)^{0.65}}$$ in which I=intensity in inch/hr and T_d= duration in minutes.

A culvert is installed at the sag to pass storm runoff across the highway. The entrance pool in front of the culvert is shaped with a slope of 5% from all directions toward a highway. The crossing culvert is designed to release an average flow of 7.50 cfs before the entrance pool becomes full. Determine the maximum water depth at the entrance pool before the highway is overtopped.

Solution: For this case, k= 0.5 or a half circle is the cross sectional geometry at the entrance pool. Let H_c= 1.0 foot. Try T_m= 16.7 min to find the maximal water depth of 0.8 foot.

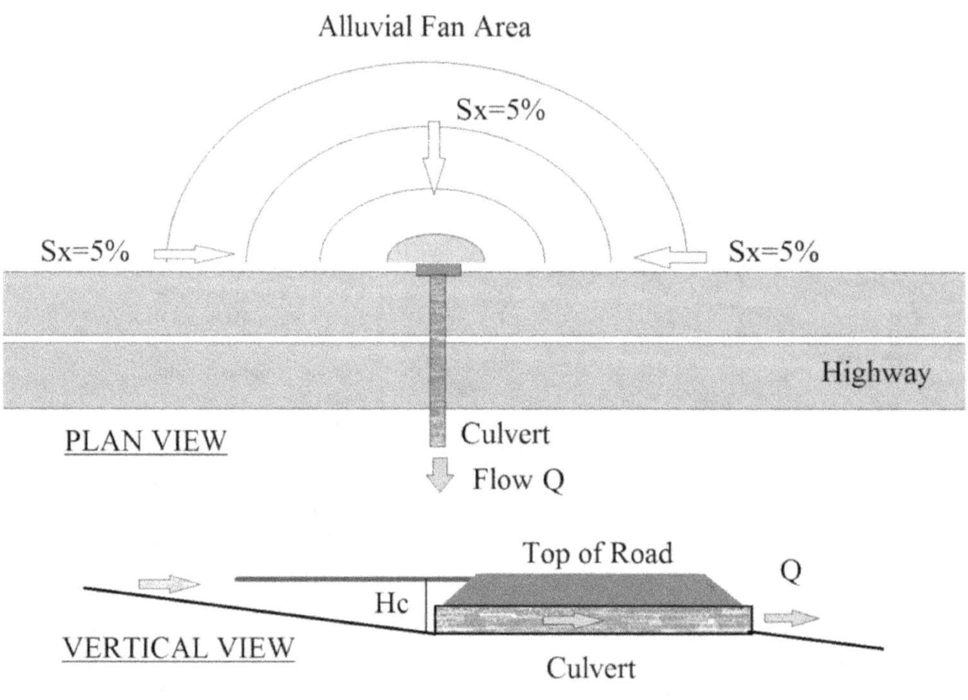

Figure HW. 5.2 Crossing Culver under Highway.

STREET DETENTIONA AND STORAGE AT SUMP					
Detention Volume for Design					
K=1 for ft or K=1/360 for meter			K=	1.00	
Tributray Area			A=	4.00	acre
Runoff Coefficient			C=	0.600	
Outflow from Sump Inlet			Q=	7.50	cfs
Rainfall Information			a=	45.50	
	$I_d = \frac{a}{(T_d + b)^m}$		b=	6.50	
			m=	0.650	
Guess Duration Tm			Tm=	16.79	minutes
Check if Tm is acceptable			Balance=	0.000	
Design Rainfall Intensity Td=Tm			I=	5.88	inch/hr
Stormwater Detention Volume			V-design=	6660.07	cubic ft
Storage Volume in Sump					
Street Transverse Slope			Sx=	0.05	ft/ft
Curb Height			Hc=	1	foot
Incremental Depth			Dy=	0.2	foot
Depth		STORAGE	VOLUME		
h (ft)		Vh	cubic ft		
k=	0.25	0.50	0.75	1.00	
0.00	0.00	0.00	0.00	0.00	
0.20	20.94	41.89	62.83	83.78	
0.40	167.55	335.10	502.66	670.21	
0.60	1518.44	2083.93	2649.42	3214.90	
0.80	5317.96	6643.36	7968.75	9294.14	

5.5. REFERENCES

Guo, James C.Y. (2000a). "*Street Storm Water Conveyance Capacity*." ASCE J. of Irrigation and Drainage Engineering, Vol 126, No. 2, Mar/Apr.

Guo, James C.Y. (2000b). "*Street Storm Water Storage Capacity* " Water Environmental Research Journal, Vol 72, No. 5, Setember/Octoer.

Guo, James C.Y. and McKenzie, K, and Mommandi, A. (2008). "*Sump Inlet Hydraulics*", ASCE J. of Hydraulic Engineering, Vol 135, No 1, Nov.

Guo, James C.Y., (1997). "*Street Hydraulics and Inlet Sizing*", published by Water Resources Publications, LLC, Littleton, Colorado

HEC-1 (2005). "*Hydrologic Package for Watershed Analysis*", Corp of Engineers, Hydrologic Engineering Center, Davis, California

CHAPTER 6
INLET HYDRAULICS

6.1. TYPES OF INLET

There are four major types of inlets used for street drainage. They are *grate inlet*, *curb opening inlet*, *combination inlet*, and *slotted inlet*. Each one has its pros and cons in practice. In general, the major concerns in the selection of street inlets are debris clogging, hydraulic efficiency, traffic interference, and public safety.

6.1.1. Grate Inlet

A grate inlet unit is composed of an inlet box with its opening covered by a grate. Grate inlets are often installed horizontally along the flow line in a street gutter. Storm water falls into the inlet box through the grate and then drains into the downstream storm sewers. Grates are formed with steel bars as shown in Figure 6.1(a) or vanes in Figure 6.1(b). A grate is described by its length, L, and width, W, type of bars/vanes, and area opening ratio after subtracting the bar and vanes areas from the total surface area.

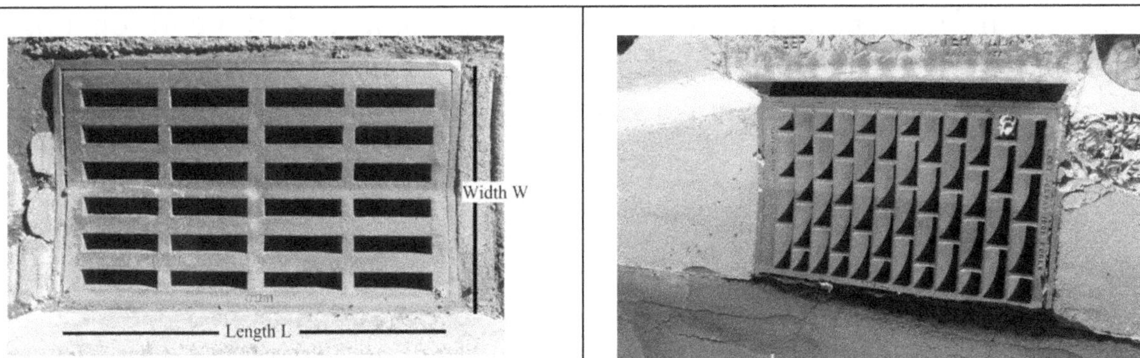

(a) Bar Grate | (b) Vane Grate

Figure 6.1. Grate Inlet.

6.1.2. Curb-opening Inlet

A unit of curb-opening inlet is composed of a vertical opening on the street curb, a depression pan within the street gutter, and an underground inlet box to transfer storm water to the downstream storm sewers. As shown in Figure 6.2a, a curb-opening unit is described by the height of the opening, H, the length of the opening, L, and the wing width, W_p, of the depression pan. Although a depression pan can enhance the hydraulic efficiency, it does interfere with bikes, snow plow, etc. Therefore, a depression pan is optional to a curb opening inlet. In comparison, a curb-opening inlet has a high hydraulic efficiency and is the least susceptible to debris clogging.

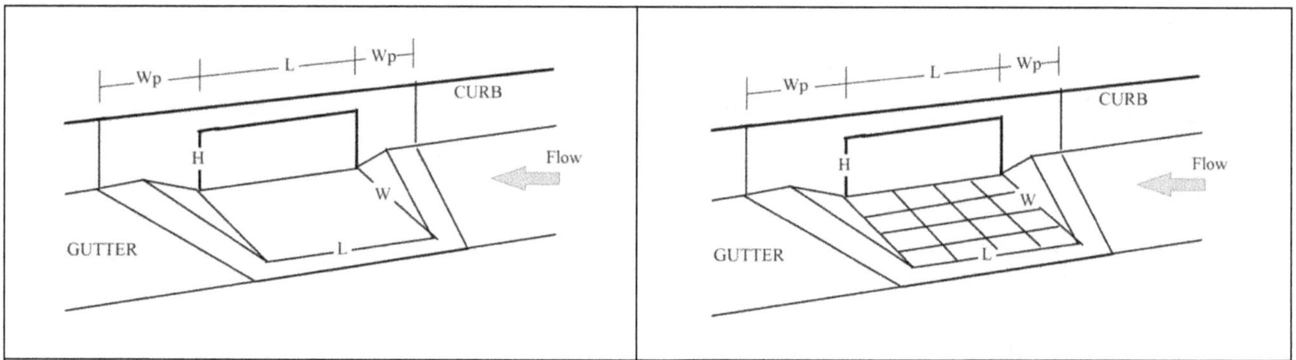

(a) Curb Opening | (b) Combo – Opening and Grate

Figure 6.2. Curb Opening Inlet.

6.1.3. Combination Inlet

As shown in Figure 6.2b, a combination inlet is composed of a curb opening inlet and a grate inlet along the flow line within the street gutter. A combination inlet shall be installed to the places where a high storm water interception is required or a high risk of debris clogging is expected.

6.1.4. Slotted Inlet

A slotted inlet in Figure 6.3 is a strip of horizontal grate placed on top of an underground trench. A slotted inlet can be either placed in parallel to the curb or perpendicular to the curb, depending on storm water spread, traffic interference, snow plowing, and other maintenance considerations.

(a) Slotted Inlet

(b) Trench Inlet

Figure 6.3. Slotted Inlet.

6.2. INLET HYDRAULICS

From hydraulic point of view, inlets are classified into two categories: (1) on a continuous grade, and (2) in a sump. Sumps are formed by the local topography or street curbs and crowns. For instance, at a street interception, the street crowns may confine the flow to the street corner. On a continuous grade, the inlet is designed to capture 70 to 80% of the flow on the street because the significant diminishing collection capacity when adding more inlets on a sloping street. At a sump, the inlet is designed to capture 100% of the peak flow under the design headwater depth or the available; whichever is less. Figure 6.4 illustrates how to identify the hydraulic condition at various inlet locations. Figure 6.5 are examples observed in field.

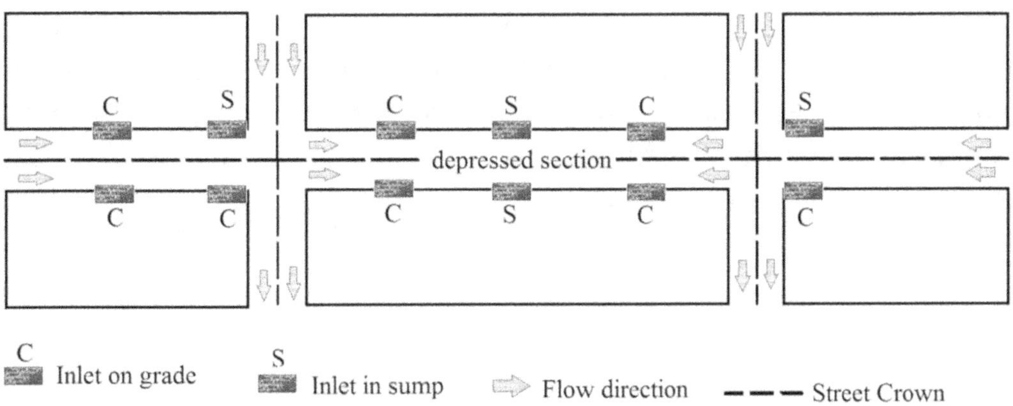

Figure 6.4. Inlets on a Grade versus Inlets in a Sump.

(a) Inlet in Sump	(b) Inlet on Grade

Figure 6.5 .Inlets on Grade and in Sump.

6.3. DETERMINATION OF DESIGN DISCHARGE

On a continuous grade, an inlet is designed to capture 70 to 80% of the design flow on the street. At the design point, the design flow consists of the local flow and the carry-over flow from the upstream inlet(s) as:

$$Q_S = Q_B + Q_{CO} \tag{6.1}$$

in which Q_S = design flow on the street in $[L^3/T]$, Q_B = local flow in $[L^3/T]$, and Q_{CO} = carryover flow in $[L^3/T]$.

<u>**EXAMPLE 6.1**</u> The catchment in Figure 6.6 is the one studied in EXAMPLE 3.2 in Chapter 3. The flow was assumed to take the flow path from Points A to B and then from B to C. The 5-year peak discharge at Point C was predicted to be 15.3 cfs. Street BC was also evaluated in EXAMPLE 4.3 to have an allowable flow capacity of 10.16 cfs. Knowing that the carry-over flow at Point B is 1.3 cfs, determine the design flow on Street BC.

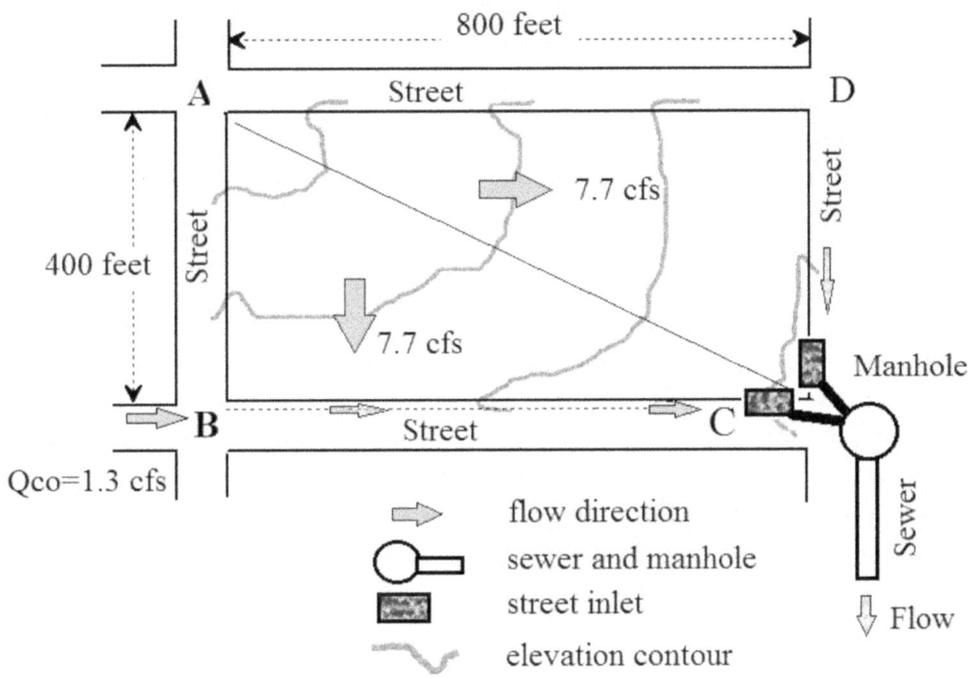

Figure 6.6. Determination of Design Flow via Street BC.

Solution

For this case, the peak flow produced from the catchment is greater than the allowable flow capacity on the street. The strategy is to split flow of 15.3 cfs into two halves. The first half of 7.7 cfs will be carried through Street BC and the second half will take Street DC. At Point C, two inlets will be installed to collect two flows from Streets BC and DC.

The design flow, Q_S, for Street BC is the sum of the two flows: Q_{CO}=1.3 cfs bypassed from the upstream inlet of Point B and the local flow, Q_B =7.7 cfs as:

Q_S=7.7+1.3=9.0< 10.16 cfs which is the allowable flow capacity

With a longitudinal slope of 1.5%, the design flow on Street BC is analyzed in Table 6.1 as:

Table 6.1. Design Flow on Street BC.

Design Flow on the Street	Qs=	**9.00**	cfs
Guess Water Spread Width <Tm	T=	14.24	ft
Gutter Cross Slope: Sw = Sx + Ds/W =	Sw	0.1033	ft/ft
Water depth Y = T*Sx =	Y=	0.28	ft
Gutter Water Depth at Curb (<Dm): D=Y+Ds =	D=	0.45	ft
Spread for Side Flow: Tx =T - W=	Tx=	12.24	ft
Spread for Gutter Flow: Ts = D/Sw =	Ts=	4.37	ft
Qws = 0.56/n* Sw^1.67* Ts^2.67*So^0.5=	Qws=	4.96	cfs
Qww = 0.56/n*Sw ^1.67*(Ts-W)^2.67*So^0.5=	Qww=	0.97	cfs
Gutter Flow: Qw=Qws-Qww =	Qw=	4.00	cfs
Side Flow: Qx = 0.56/n* Sx^1.67* Tx^2.67*So^0.5=	Qx=	5.01	cfs
Calculated Total Flow: Qs = Qx + Qw =	Qs=	9.00	cfs
Flow Difference to seek the solution	dQ=0	0.00	cfs
Check on the VD product			
Flow Area As =0.5*(Ytx+DsW)=	As=	1.91	sq ft
Flow Velocity Vs = Qs/As =	Vs=	4.71	fps
The VD product = VsD =	VsD=	2.13	

6.4. CLOGGING FACTOR

It is recommended that a clogging factor of 50% be applied to a single grate, and 10% be applied to a single curb opening unit. As illustrated in Figure 6.7, for a series of inlets, it is reasonable to assume that the clogging percentage decays as the number of inlet increases.

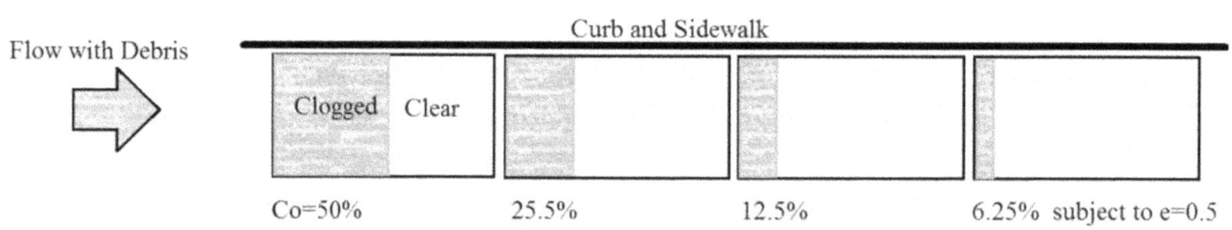

Figure 6.7. Decay of Inlet Clogging Percentage.

As a result, the clogging factor for multiple inlets in serial is equal to the total clogging percentage divided by the number of inlet units as (Guo 2000c, Guo 2006):

$$C_g = \frac{1}{N}\left(C_o + eC_o + e^2C_o + e^3C_o + \ldots\ldots\ldots + e^{N-1}C_o\right) = \frac{C_o}{N}\sum_{i=1}^{i=N} e^{i-1} \qquad (6.2)$$

in which C_g = multiple-unit clogging factor, $0 \leq C_g \leq 1$, C_o = single-unit clogging factor, e = decay ratio less than unity, and N = number of inlets. Field data suggested e=0.25 to 0.5. When N becomes large, Eq 6.2 is converged to:

$$C_g = \frac{C_o}{N(1-e)} \tag{6.3}$$

The interception of an inlet on a grade is proportional to the inlet length, and in a sump is proportional to the inlet opening area. Therefore, a clogging factor shall be applied to the length of an inlet on a continuous grade as:

$$L_e = (1-C_g)L \tag{6.4}$$

in which L_e = effective (unclogged) length in [L]. Similarly, a clogging factor shall be applied to the opening area of an inlet in a sump as:

$$A_e = (1-C_g)A \tag{6.5}$$

in which A = opening area in [L^2], and A_e = unclogged opening area in [L^2].

6.5. GRATE INLET ON A CONTINUOUS GRADE

As illustrated in Figure 6.8, the storm water moving on a street can be divided into the *gutter flow* that is carried by the street gutter within the gutter width, and the *side flow* that is the water spread into the traffic lanes.

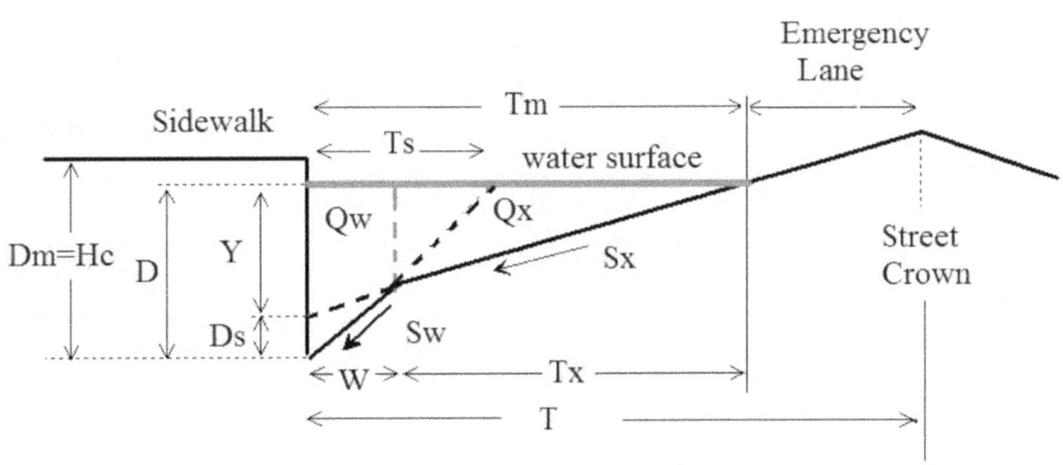

Figure 6.8. Parameters for Street Flow.

The interception of gutter flow by a grate is determined by the length of the grate, average cross sectional water velocity, and water splash velocity due to the interference of the grate. A regression analysis using the laboratory data reported by HEC-12 results in the following empirical formula for determining the splash-over velocity, V_o, as a function of the grate length and type.

$$V_o = \alpha + \beta L_e - \gamma L_e^2 + \eta L_e^3 \tag{6.6}$$

in which V_o = water splash velocity in feet/second on grate, L_e = effective length in feet, α, β, γ, and η = constants, depending on the type of grate as shown in Table 6.2.

Table 6.2. Coefficients to Determine Splash Velocity over Inlet Grates.

Type of Grate	α	β	γ	η
Bar P-1-7/8	2.22	4.03	0.65	0.06
Bar P-1-1/8	1.76	3.12	0.45	0.03
Vane Grate	0.30	4.85	1.31	0.15
45-Degree Bar	0.99	2.64	0.36	0.03
Bar P-1-7/8-4	0.74	2.44	0.27	0.02
30-Degree Bar	0.51	2.34	0.20	0.01
Reticuline	0.28	2.28	0.18	0.01

(a) Shallow Water Splash Velocity | (b) Splash Velocity Over Grate

Figure 6.9. Splash Velocity over Grate Observed in Field.

The interception capacity of a grate is separately determined for the gutter and side flows. The interception percentage of a gutter flow, Q_w, is estimated as:

$$R_w = 1 - 0.09(V_s - V_o) \text{ if } V_s \le V_o \text{, otherwise } R_w = 1 \quad (6.7)$$

in which R_w = interception ratio of gutter flow, and V_s = average cross sectional water velocity in feet/second which can be obtained from the street runoff flow analysis. For most cases, the condition, $V_s<V_o$, prevails, or R_w =1.0. The interception percentage of the side flow, Q_x, is expressed by

$$R_x = \frac{1}{\left(1 + \dfrac{K_R V^{1.8}}{S_x L_e^{2.3}}\right)} \quad (6.8)$$

in which R_x = interception ratio of side flow, K_R=0.15 for feet-second units or 0.083 for meter-second units. As a result, the total interception capacity, Q_a, for a grate inlet is equal to

$$Q_a = R_w Q_w + R_x Q_x \quad (6.9)$$

EXAMPLE *6.2* A unit of bar grate has a width of 1.58 feet and a length of 2.96 feet. Considering a clogging factor of 0.5 for a single unit, determine the number of bar grates required to intercept approximately 80% of the design flow in EXAMPLE 6.1.

Referring to Table 6.1, we have Q_w= *4 cfs*, Q_x=*5 cfs*, and V_s=*4.71 feet/sec.* Consider four bar grates. The total grate length is:

$$L_g = nL = 4 \times 2.96 = 11.84 \text{ ft}$$

The clogging factor is 0.5 for a single unit. With e= 0.25, the clogging factor for four units is computed by Eq 6.3 as:

$$C_g = 0.5/[4x(1-0.25)] = 0.17$$

The effective grate length free from clogging is:

$$L_e = (1-0.17)x11.84 = 9.87 \text{ ft}$$

Assuming that the empirical coefficients developed for Bar P-1-7/8 are suitable for the Bar Grate in this case, we have

$$V_o = 2.22 + 4.03L_e - 0.65L_e^{\ 2} + 0.06L_e^{\ 3} = 36.3 \ \textit{feet/second} > V_s, \ \ \textit{So, } R_f = 1.0$$

$$R_x = \frac{1}{1 + \dfrac{0.15 \times 4.71^{1.8}}{0.02 \times 9.87^{2.3}}} = 0.61$$

The interception is calculated as:

$$Q_c = 1.0x4.0 + 0.61x5.0 = 7.1 \ cfs$$

Using four units, the interception ratio for this example is: 7.1/9.0 = 79.0%

6.6. GRATE INLET IN A SUMP

A grate inlet in a sump operates like a weir under a shallow depth. Its capacity is estimated as:

$$Q_w = \frac{2}{3} C_d \sqrt{2g} P_e D^{1.5} = C_w P_e D^{1.5} \tag{6.10}$$

in which Q_o = interception capacity in $[L^3/T]$, C_d= discharge coefficient such as 0.6, C_w = weir coefficient such as 3.0 for feet-second units or 1.77 for meter-second, Y_s = water depth in [L], and P_e = effective weir length in [L]. As illustrated in Figure 6.10, water flows may overtop the clogged areas from the sides. As a result, the effective wetted parameter for a weir flow is estimated as:

$$P_e = m[(1 - C_g)L + 2W] \tag{6.11}$$

Where m=length opening ratio, such as 0.7, after subtracting the length of steel bars.

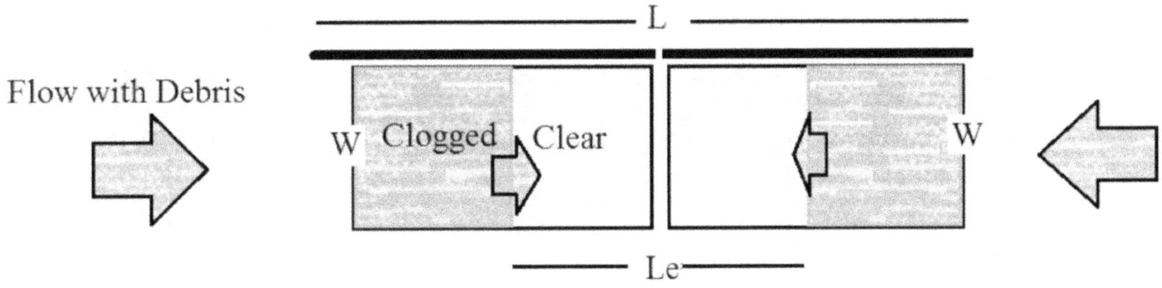

Figure 6.10. Effective Wetted Parameter for Clogged Grates.

When a submerged grate operates like an orifice, its capacity is estimated as:

$$Q_o = C_d A_e \sqrt{2gY_s} \tag{6.12}$$

$$A_e = n(1 - C_g)WL \tag{6.13}$$

in which C_d = discharge coefficient for orifice flow such as 0.60, g = gravitational acceleration in [L/T^2], W= grate width in [L], L=grate length in [L], and n = area opening ratio, such as 0.5 to 0.6, on the grate after subtracting the area of steel bars. Transition between weir flow and orifice flow is not clearly understood. In practice, for a specified water depth on a grate, the interception capacity is evaluated by both Eq 6.10 and 6.12, and the smaller one dictates.

$$Q_a = \min(Q_w,\ Q_o) \tag{6.14}$$

EXAMPLE 6.3 A bar grate has a length of 2.96 ft and width of 1.58 ft. The length opening ratio m=0.7 while the area opening ratio n=0.45. The clogging factor is C_o=0.50 for a single unit and decayed with e=0.25. Determine the number of grates for Street BC under a water depth of 0.5 foot.

Design Discharge on the Street	Qs=	9.00	cfs
Length of Unit Grate	L=	2.96	ft
Width of Unit Grate	W=	1.58	ft
Clogging Factor for Single Grate	Co=	0.50	
Number of Unit Grate	N=	4.00	
Clogging Factor for Grate Inlet	Cg=	0.17	
Effective Length of Grate Inlet = (1-Clog)*L=	Le=	9.87	ft
Weir Coefficient	Cw=	3.00	
Discharge Coefficient for Orifice Flow	Cd=	0.60	
Guess Water Depth <Dm	Yd=	0.48	ft
As a Weir			
Length Opening Ratio of the Grate	m=	0.70	
Capacity as a Weir=mCw(Le+2W)Yd^1.5=	Qweir=	9.00	cfs
As an Orifice			
Areal Opening Ratio of the Grate	n=	0.45	
Capacity as an Orifice =Co(n Le W) (64.4*Yd)^0.5=	Qorifice=	23.32	cfs
Capacity for Design=min(Qweir, Qorifice)	*Qa=*	9.00	cfs
Check if ΔQ=0 because of no carry-over flow	Q-co=	0.00	cfs
Is Yd <Dm ?	Check	YES	

Using four (4) grates, the ponding depth was found to be 0.48 foot<0.5 foot (Ok!) $Q_a=min(Q_w, Q_o)=min\ (9.0, 23.32)=9.0\ cfs$. Field observations of weir ad orifice flows at a grate inlet are presented in Figure 6.11.

(a) Orifice Flow on Grate

(b) Weir Flow around Grate

Figure 6.11. Weir and Orifice Flows at Grate.

6.7. CURB OPENING ON A GRADE

To install a curb opening inlet on a continuous grade, the required curb opening length, L_t, for a complete interception of the design storm runoff, Q_s, on the street is computed by:

$$L_t = K_t Q^{0.42} S_o^{0.30} \left(\frac{1}{nS_e} \right)^{0.6} \quad (6.15)$$

$$S_e = S_x + S_w \frac{Q_w}{Q_s} \quad (6.16)$$

in which L_t = required length for a 100% runoff interception, K_t=0.6 for feet-second units or 0.82 for meter-second units, S_o = street longitudinal slope in [L/L], n = Manning's roughness of 0.016, and S_e = equivalent transverse street slope in [L/L]. The curb-opening inlet shall have a length less than, but close to, L_t. The interception capacity of a curb-opening inlet is calculated as:

$$Q_a = Q\left[1 - \left(1 - \frac{L_e}{L_t}\right)^{1.80}\right] \quad (6.17)$$

in which Q_a = inlet capacity in [L^3/T], and L_e = effective length in feet of curb opening inlet.

EXAMPLE 6.4 A unit of curb opening inlet has a length of five feet. Considering a clogging factor Co= 0.12 for a single unit and e=0.25 for debris decay. Determine the interception rate to apply three (3) units to the design flow in EXAMPLE 6.1.

Design Discharge on the Street	*Qs*=	9.00	cfs
Length of Unit Curb Opening	*L*=	5.00	ft
Clogging Factor for Single Curb Opening	*Co*=	0.10	
Number of Unit Curb-opening	*N*=	3.00	
Clogging Factor for Curb-opening Inlet	*Cg*=	0.04	
Unclogged Length Le = (1-Clog)L =	*Le*=	14.33	ft
Gutter Flow to Design Flow Ratio=Qw/Qs=	*Eo*=	0.44	ft
Equivalent Slope Se = Sx + Sw.Eo =	*Se*=	0.066	ft/ft
Length Lo =0.60*Qo^0.42*So^0.30*[1 /(nSe)]^0.6=	*Lo*=	26.19	ft
Interception Capacity = Qs(1-(1-L/Lo)^1.8)=	*Qa*=	6.84	cfs
Carryover flow = Qs-Qa=	*Q-co*=	2.16	cfs
Capture percentage for this inlet = Qa/Qs =	*C%*=	76.0%	

The flow interception rate for three cub-opening units is 76%.

6.8. CURB-OPENING INLET IN A SUMP

Referring to Figure 6.2, a curb-opening inlet in a sump operates like a weir. Its interception capacity is estimated as:

$$Q_w = \frac{2}{3} C_d \sqrt{2g} P_e Y_s^{1.5} = C_w P_e Y_s^{1.5} \tag{6.26}$$

$$P_e = (1 - C)(L + kW_p) + 2W \tag{6.27}$$

in which P_e = effective weir length in [L] around the depressed pan in front of curb opening inlet, W_p= width of depressed pan in [L], and k = 1.8 to 2.0 for two sides of the pan in Figure 6.2. When the water depth gets deeper, a curb opening inlet operates like an orifice and can be modeled as:

$$Q_o = C_d A_e \sqrt{2g(Y_s - 0.5H)} \tag{6.28}$$

$$A_e = (1 - C)HL \tag{6.29}$$

in which H = height of curb opening in [L]. The transition of an inlet from weir flow to orifice flow is not clearly defined. In practice, both Eq 6.26 and 6.28 shall be applied to the water depth and the smaller one shall be used for the inlet interception.

EXAMPLE 6.5 Consider a clogging factor of 12%, and the debris decay factor of 0.25. The maximum water depth is 0.5 foot at the curb face. Determine the number of the curb-opening units in Figure 6.12 for the flow in EXAMPLE 6.1.

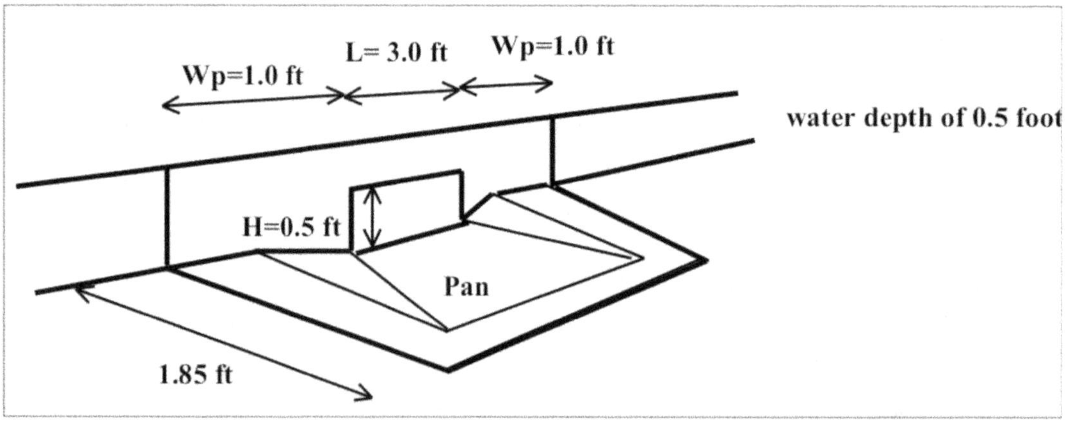

Figure 6.12. Example of Curb Opening Inlet in Sump.

Design Discharge on the Street	Qs=	9.00	cfs
Height of the Curb Opening	H=	0.50	ft
Length of Unit Curb Opening	L=	3.00	ft
Clogging Factor for Single Curb Opening	Co=	0.10	
Number of Unit Curb-opening	N=	3.00	
Clogging Factor for Grate Inlet	Cg=	0.04	
Length of the Curb Opening Inlet	NL=	9.00	ft
Width of Depressed Pan Or Gutter	W=	1.85	ft
Side Width for Depression Pan	Wp=	1.00	ft
Weir Coefficient	Cw=	3.00	
Discharge Coefficient for Orifice Flow	Cd=	0.65	ft
Guess Water Depth<Dm	Yd=	0.41	ft
As a Weir			
Capacity as a Weir =Cw[(1-Clog)(NL+2Wp)+2W)]Yd^1.5=	Qweir=	11.23	cfs
As an Orifice			
Capacity as an Orifice=Co(1-Clog)H(NL)(64.4(Yd-0.5H))^0.5=	Qorifice=	9.00	cfs
Capacity for Design=min(Qweir, Qorifice)	*Qa*=	9.00	cfs
Check if ΔQ=0 because of no carry-over flow	ΔQ=	0.00	cfs
Is Yd <Dm ?	Check	YES	

For this case, it takes three (3) units under a ponding depth of 0.41 foot<0.5 foot to drain the design flow of 9.0 cfs.

(a) Shallow water at Sump | (b) Deep Water at Sump

Figure 6.13. Flow at Sump Curb Opening Inlet.

6.9. SLOTTED INLET

Hydraulically, a slotted inlet is similar to a curb-opening inlet. As a result, design formulas developed for curb-opening inlet are also applicable to slotted drain inlets.

6.10. COMBINATION INLET

A combination inlet in Figure 6.14 consists of a curb opening and a grate inlet. During the storm event, if one is clogged, as shown in Figure 6.14, the other can still function. Empirical formulas for sizing a grate or a curb opening were developed under the condition that the inlet operates independently. Interference between the grate and the curb opening in a combination inlet has not been fully understood yet. The assumption of independent operations implies that the curb-opening inlet is placed immediately downstream of the grate inlet. In other words, the curb-opening inlet receives the carryover flow from the grate inlet. In case of a 100% interception by the grate, the curb opening inlet will intercept no flow at all.

In theory, the capacity of a combination inlet is the sum of the intercepted discharges by the grate and the curb opening. To be conservative, the capacity of a combination inlet is the higher between the grate and the curb-opening (Guo, 1997)(Guo, McKenzie and Mommandi 2008).

(a) Clogged Grate

(b) Clogged Curb-opening

Figure 6.14. Clogged Combination Inlet.

6.11. CARRYOVER FLOW

When the street flow on a grade is not completely collected into a street inlet, the residual flow is carried to the downstream inlet. The carryover flow is defined as:

$$Q_c = Q_s - Q_a \qquad (6.30)$$

in which Q_c = carry-over flow. At a sump, the flow is supposed to be completely intercepted, except that a crossing concrete pan in Figure 6.15 is available to bypass the excess water flow.

(a) Carryover Flow	(b) Crossing Pan

Figure 6.15. Crossing Concrete.

The carryover flow must be treated as a part of the design flow for the immediately downstream inlet. As aforementioned, a carryover flow can be directly added to the local flow.

6.12. CLOSING

On a grade, the flow interception at an inlet is sensitive to the street longitudinal slope. The diminishing return on adding more inlets discourages a high flow interception rate. Therefore it is important in practice that an inlet in grade be designed to capture 70 to 80% of the flow on the street. Obviously, a sump inlet has to be designed to have a complete interception within the gutter width. An undersized inlet at sump will inundate the adjacent sidewalks. Flow interception at an inlet is dominated by either weir or orifice flow, depending on the flow depth. On top of the hydraulic complexity, the inlet flow is further complicated by debris clogging. The design procedures discussed in this Chapter are semi-theoretical. As expected, more research efforts are required to improve the understanding of the mixing flow between orifice and weir flows.

6.13. HOMEWORK

Q 6.1 Street inlets are installed at the street corner as shown in Figure HW 6.1. The design information is given as follows:

(a) The 5-yr design rainfall intensity, I (inch/hr), is given as:

$I(inch/hr)=\dfrac{28.5P_1}{(10+T_d)^{0.789}}$ in which P_1 = 1.35 inch and T_d=duration in minutes.

(b)The overland flow time, T_o, shall be calculated by the airport formula for a flow length ≤300 feet. The swale and gutter flow velocities, V_f, are estimated by the SCS upland method with a conveyance coefficient of 20.

Your tasks are:

a) Inlet 1 is placed on grade because it has a downstream crossing pan. Size Inlet 1 using grate inlets.

b) Inlet 2 is placed on grade for it has a downstream pan. Size Inlet 2 using curb opening inlets.

c) Inlet 3 is in sump. Size Inlet 3 using combo inlets.

d) Inlet 4 is in sump. Size Inlet 4 using combo inlets.

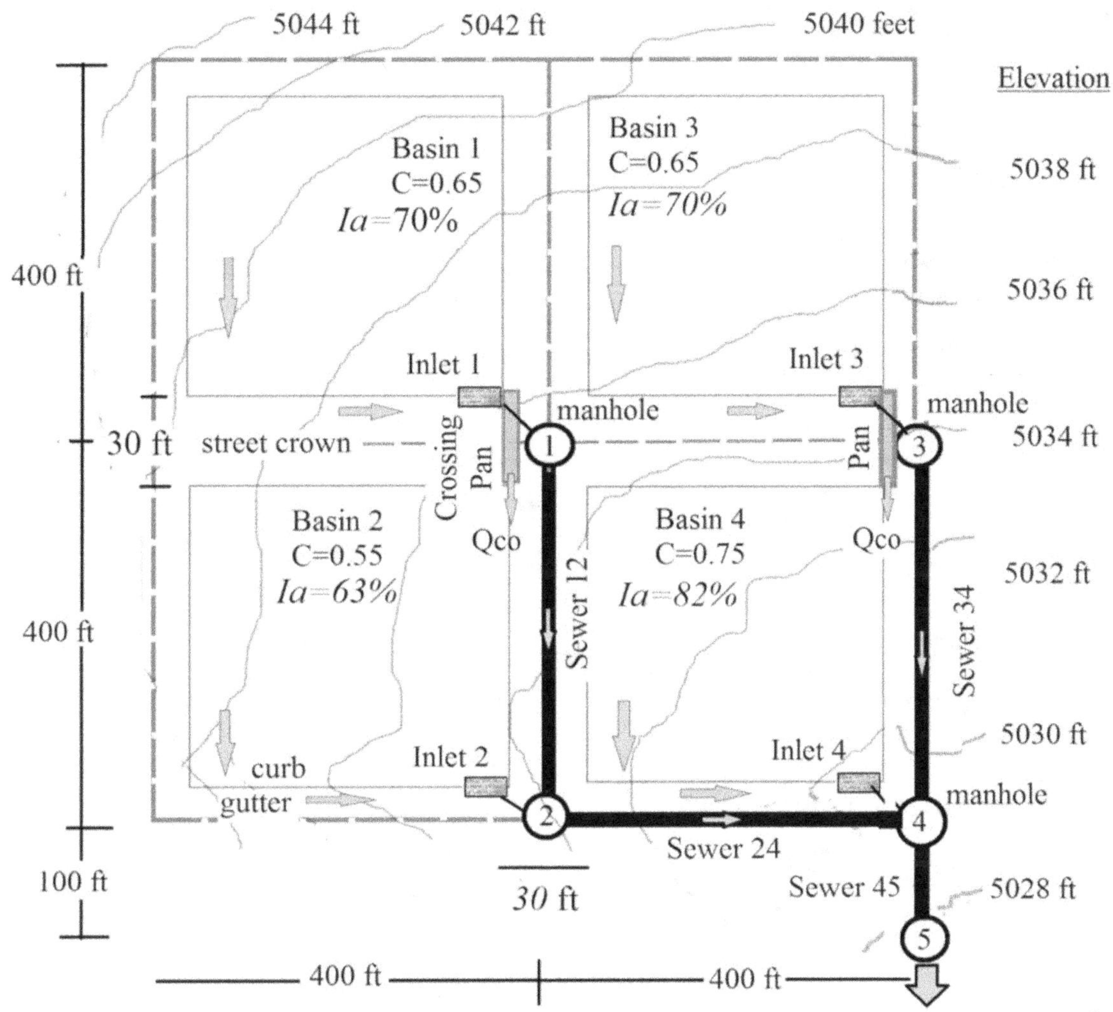

Figure HW 6.1. Design of Street Inlets.

Solution for Inlet 1

Step 1: Design Flow

5-yr Coefficient	Overland Flow			Swale Flow				Gutter Flow				Existing Tc
	Slope	Length	Time	Slope	Length	SCS K	Time	Slope	Length	SCS K	Time	
	%	*ft*	*min*	%	*ft*	*fps*	*min*	%	*ft*	*fps*	*min*	*min*
C5	*So*	*Lo*	*To*	*S2*	*L2*	*K2*	*T2*	*S3*	*L3*	*K3*	*T3*	*Tc-comp*
0.65	1.00	300.00	14.07	1.00	100.00	20.00	0.83	1.00	400.00	20.00	3.33	18.24

Drainage Area	Imp Ratio	Design Runoff Coef	Total Flow Length	Average Slope	Initial Time	Convey Factor	Flow Velocity	Flow Time	Future Tc	Design Storm Duration	Rainfall Intensity 5-yr	Peak Flow
acre		*C-design*	*ft*	%	*min*	*fps*	*fps*	*min*	*min*	*min*	*in/hr*	*cfs*
A	*Ia*		*L*	*S*	*T**	*K**	*V**	*Tf*	*Tc-Reg*	*Td*	*I*	*Qp*
3.67	0.70	0.65	800.00	1.00	7.50	28.80	2.88	4.63	12.13	12.13	3.34	7.98

Step 2: Design Flow on the Street

A. Street Geometry				
Curb Height	H=	6.00	*inches*	
Gutter WIdth	W=	2.00	*ft*	
Gutter Depression	Ds =	2.00	*inches*	
Street Transverse Slope	Sx=	0.0200	*ft/ft*	
Street Longitudinal Slope	So=	0.0100	*ft/ft*	
Manning's Roughness	n=	0.0160		
Design Discharge on the Street	Qs=	**7.98**	*cfs*	
B. Street Conveynace Capacity				
Guess Water Spread Width for Design Qs	T=	14.75	*ft*	**(Guess)**
Water depth Y = T*Sx =	Y=	0.30	*ft*	
Water Depth at Curb: D=Y+Ds =	D=	0.46	*ft*	
Spread for Side Flow: Tx =T - W=	Tx=	12.75	*ft*	
Gutter Cross Slope: Sw = Sx + Ds/W =	Sw	0.103	*ft/ft*	
Spread for Gutter Flow: Ts = D/Sw =	Ts=	4.47	*ft*	
Gutter Flow	Qw=	3.42	*cfs*	
Side Flow	Qx=	**4.56**	*cfs*	
Total Flow from the Guessed T	Qs=	**7.98**	*cfs*	
Check on Design Flow Q=0	Q=	0.00	*cfs*	**(Check)**
Flow Area As = 0.5YT+0.5WDs =	As=	2.34	*sq ft*	
Flow Velocity Vs = Q/As =	Vs=	3.41	*fps*	
The VD product VsD =	VD=	**1.57**	*ft^2/s*	

Step 3: Size Inlet 1 using Grates

Design Discharge on the Street	*Qs=*	7.98	*cfs*
Length of Unit Grate	*L=*	2.96	*ft*
Clogging Factor for Single Grate	*Co=*	0.50	
Number of Unit Grate	*N=*	4.00	
Clogging Factor for Grate Inlet	*Cg=*	0.17	
Effective Length of Grate Inlet = (1-Clog)*L=	*Le=*	9.87	*ft*
Interception Rate of Gutter Flow	*Rf=*	1.00	
Design Flow Velocity	*Vs=*	3.41	*cfs*
Interception Rate of Side Flow=1/(1+0.15Vs^1.8/(SxLe^2.3))=	*Rx=*	0.74	
Interception Capacity=Rf*Qw+RxQx=	*Qa=*	6.79	*cfs*
Carry-over Flow = Qs-Qa =	*Q-co=*	1.19	*cfs*
Capture percentage = Qa/Qs=	*C%=*	85.1%	

Repeat the above procedure for Inlet 2, 3, and 4.

6.14. REFERENCES

Guo, James C.Y. (2000a). "*Street Storm Water Conveyance Capacity*." ASCE J. of Irrigation and Drainage Engineering, Vol 126, No. 2, Mar/Apr.

Guo, James C.Y. (2000b). "*Street Storm Water Storage Capacity* " Water Environmental Research Journal, Vol 72, No. 5, Setember/Octoer.

Guo, James C.Y. (2000c). "*Design of Grate Inlets with Clogging Factor*," Advances in Environmental Research, Vol 4, Elsvier Science, Ireland.

Guo, James C.Y. (2006). "*Decay-based Clogging Factor for Curb Inlet Design*", Vol 132, No. 11, ASCE J. of Hydraulic Engineering, November.

Guo, James C.Y. and McKenzie, K, and Mommandi, A. (2008). "*Sump Inlet Hydraulics*", ASCE J. of Hydraulic Engineering, Vol 135, No 1, Nov.

Guo, James C.Y., (1997). "*Street Hydraulics and Inlet Sizing*", published by Water Resources Publications, LLC, Littleton, Colorado

HEC12 (1984). "*Drainage of Urban Highway Drainage*". Hydraulic Engineering Circular No. 12, US Department of Transportation, Federal Highways Administration.

HEC22 (2010). "Urban *Drainage Manual*", Hydraulic Engineering Circular No. 22, US Department of Transportation, Federal Highways Administration.

UDFCD Manual (2010), "*Storm Water Drainage Design Criteria Manual*", Vol 1 and Vol 2, Urban Drainage and Flood Control District Denver, Colorado.

CHAPTER 7
STORM SEWER SYSTEM DESIGN

7.1. LAYOUT OF STORM SEWER SYSTEM

A storm sewer system consists of a series of pipes and manholes. Surface runoff enters a manhole through street inlets. Between two adjacent manholes, the flow is carried by a pipe. Figure 7.1 is the plan view of the sewer system. The layout of a storm sewer system is governed by many factors including:

(a) street alignment for the sewer line,
(b) street inlet placement,
(c) existing utility locations,
(d) sewer system outfall location, and
(e) watershed topography tributary to the sewer system.

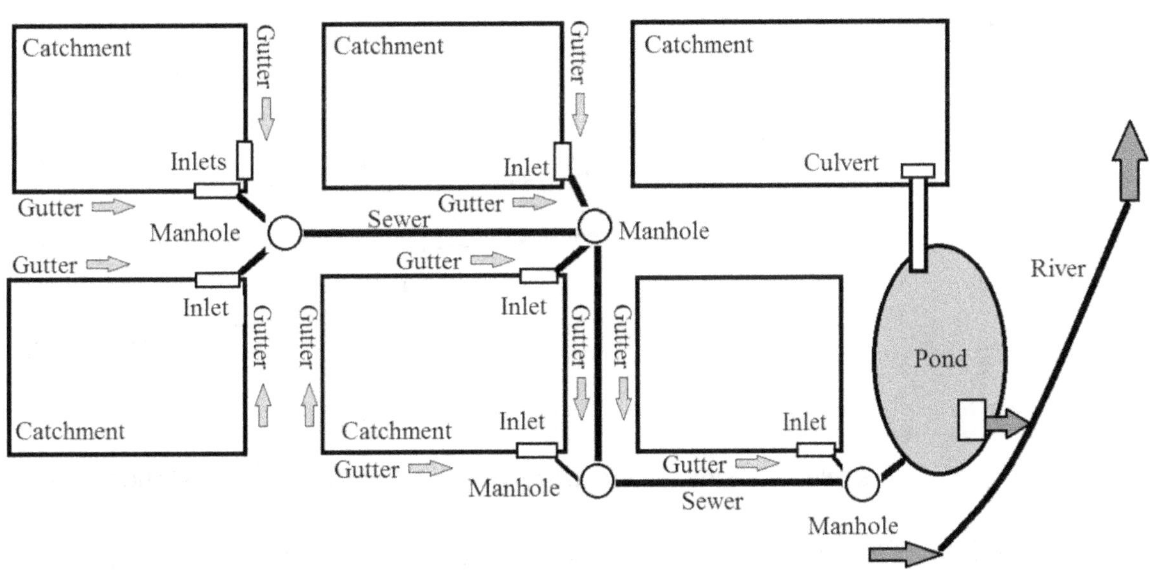

Figure 7.1. Plan View of Sewer System.

7.2. VERTICAL PROFILE

The layout of a sewer system is depicted by *plan view* and *vertical view* as illustrated in Figure 7.2. The vertical profile of a sewer line must avoid any conflict with the existing underground utilities and satisfy all design constraints. As a rule of thumb, the ground slopes provide guidance to select the first approximation of the sewer line grades. The vertical profile of a sewer system has to be adjusted several times in order to satisfy the design criteria. For instance, the slope of a sewer line must be repeatedly adjusted with manhole drops several times until the flow velocity in the sewer line is within the permissible limits, and also meet the required minimum soil cover. Sewer flows can be very complicated hydraulically. On a steep slope, the sewer carries an open channel flow. After a hydraulic jump, the sewer is surcharged. Under continuous backwater effects, the sewer may be pressurized and become flowing full. The energy grade line (EGL) and hydraulic grade line (HGL) are sensitive to the vertical profile selected for the sewer system. It takes several iterations to develop an acceptable vertical profile for the sewer system under design.

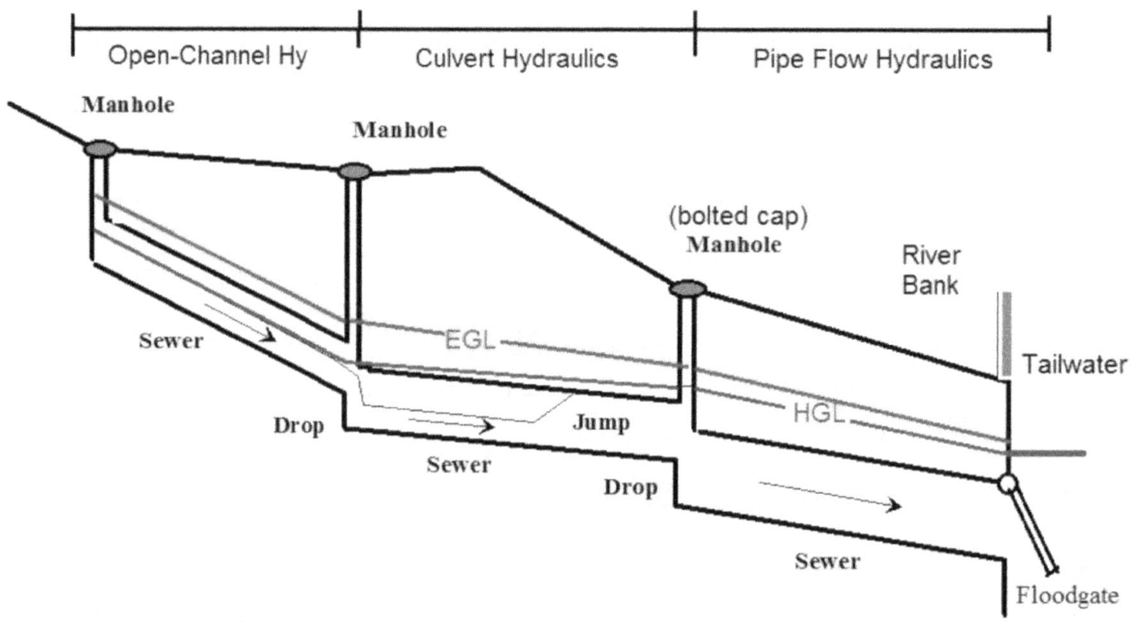

Figure 7.2. Flow Conditions in a Sewer-manhole System.

As illustrated in Figure 7.3, all manholes are also marked by its identification number such as M1, M2 etc. All sewers are marked with its length, slope, and identification number such as S12 from M1 to M2, and S23 from M 2 to M3 etc. A manhole is covered by a metal cap. The *rim elevation* of the metal cap is set to be at the local ground elevation. Each incoming sewer at the manhole is marked with its *crown* and *invert elevations*, so is the outgoing sewer. The elevation difference between the incoming and outgoing sewer inverts is termed *manhole drop*. As a common practice, it is preferred to add a floor drop of 0.20 foot at a manhole.

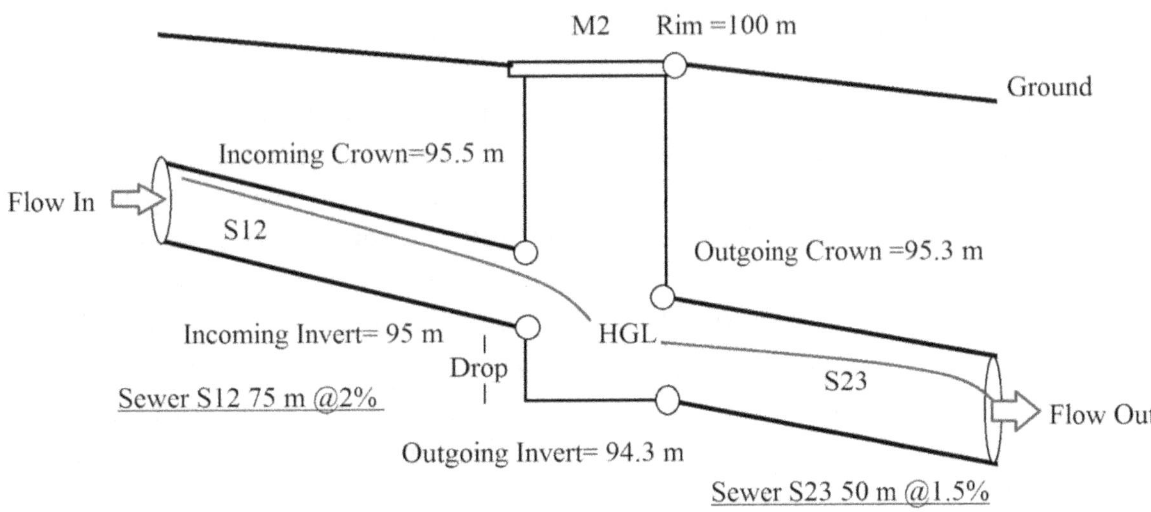

Figure 7.3. Details of Vertical Profile for Sewer System.

7.3. MANHOLE IN SEWER SYSTEM

A manhole is constructed to provide an efficient transition between incoming and outgoing sewers or to serve as an access to storm sewers for maintenance purposes. A manhole shall be placed at where:

(a) the pipe size changes,

(b) the direction of sewer line changes,

(c) the invert grades (slopes) along the sewers change,

(d) the drops are added to the sewer vertical profiles,

(e) the laterals comes to join the trunk line,

(f) the cleanout access is needed, and

(g) the spacing between manholes exceeds 400 feet.

7.4. INCOMING LATERALS

The angle of confluence between the main line and the lateral shall not exceed 45 degrees. A connector pipe from a street inlet box may join the main line at an angle greater than 45 degrees up to a maximum of 90 degrees. Care must be taken when the backwater effects from the main line may severely impact the flow conditions in the lateral. A smooth transition can significantly reduce the surcharge condition under backwater effects.

7.5. CLASSIFICATION OF SEWES

The design discharge at a manhole is determined by the cumulative tributary area. The downstream sewer at each manhole is then sized by the assumption of open channel flow under the normal flow condition. The calculated hydraulic pipe size is often not available in manufactory. As a result, the next larger commercially available pipe size shall be used. For instance, the available commercial circular pipe sizes in diameter are: 6, 8, 10, 12, 15, 18, 21, 24, 30, 33, 36, 42, 48, 54, 60, 66, 72, 78, 84-inch etc. Sewers are prefabricated with various shapes (see Figure 7.4) and materials including concrete or corrugated steel metal. If the detailed hydraulic properties are not readily available for sewers with various shapes, the equivalent circular shape is recommended for hydraulic calculations.

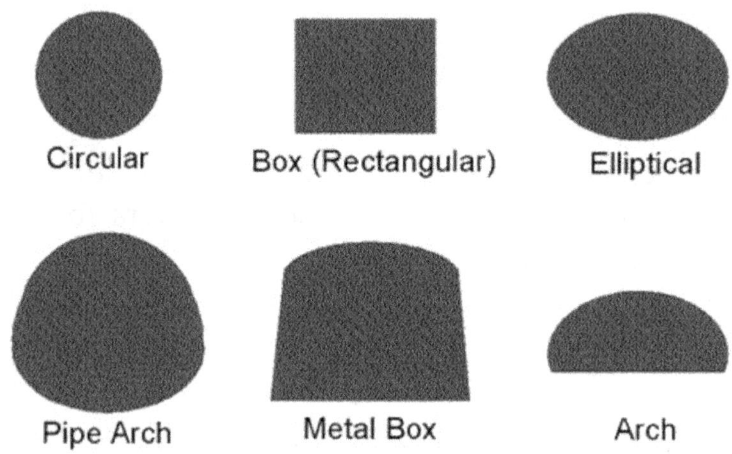

Figure 7.4. Various Sewers.

7.6. SEWER SIZING

In practice, all sewers in a drainage system are firstly sized under the assumption of open channel flow without considering downstream and upstream influences. Manning's formula for open channel flow is recommended for sewer sizing as:

$$V = \frac{K_n}{n} R^{\frac{2}{3}} \sqrt{S_s} \tag{7.1}$$

$$R = \frac{A}{P} \tag{7.2}$$

$$Q = VA \tag{7.3}$$

$$F_r = \sqrt{\frac{Q^2 T}{g A^3}} \tag{7.4}$$

in which V = cross sectional normal velocity in [L/T], K_n= 1.486 for foot-sec units or 1.0 for meter-second units, R = hydraulic radius in [L], S_s = sewer slope in [L/L], n = Manning's roughness, P = wetted perimeter in [L], A = flow area in [L^2] square feet, Q = flow rate in [L^3/T], g = gravitational acceleration in [L/T^2], T = top width of the flow area in [L], and F_r = Froude number in [$L^0M^0T^0$]. Equations 7.1 to 7.4 are further expanded into various cross sectional geometries.

7.6.1. Circular Sewer

As shown in Figure 7.5, the hydraulic parameters for a partially full flow in a circular pipe can be related to the half of central angle as:

$$y = \frac{d}{2}(1 - \cos\theta) \tag{7.5}$$

$$A = \frac{d^2}{4}(\theta - \sin\theta\cos\theta) \tag{7.6}$$

$$P = d\theta \tag{7.7}$$

$$T = d\sin\theta \tag{7.8}$$

in which θ= a half of central angle in radians, d = diameter of pipe in [L], T= top width in [L], and y = flow depth in [L].

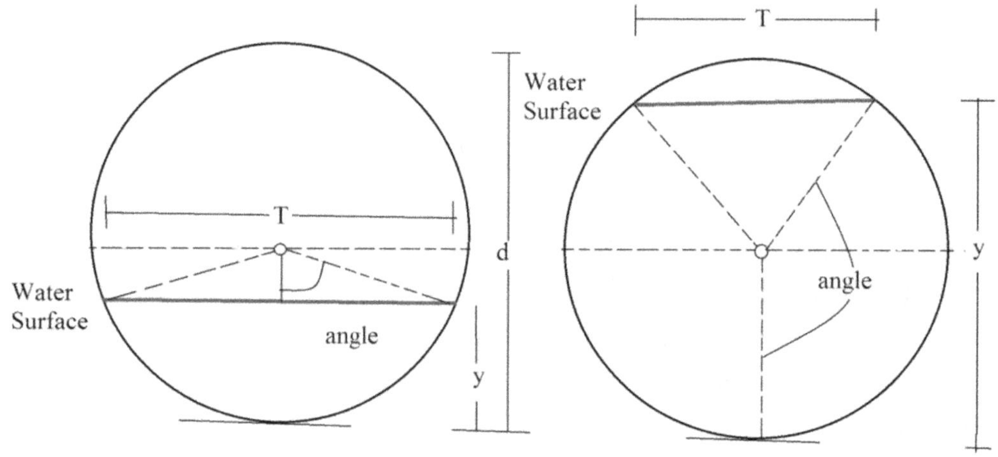

Figure 7.5. Hydraulic Parameters in a Circular Sewer.

When a circular pipe is under a flowing full condition, the half of the central angle is 180-degree. The hydraulic parameters for flowing full are calculated as:

$$A_f = \frac{\pi d^2}{4} \tag{7.9}$$

$$R_f = \frac{d}{4} \tag{7.10}$$

Substituting Eq's 7.9 and 7.10 into Eq 7.3 yields

$$d = \left(\frac{nQ}{K\sqrt{S_0}} \right)^{\frac{3}{8}} \tag{7.11}$$

Where Q = design flow in [L^3/T], S_o = conduit invert slope in [L/L], d = hydraulically required circular diameter in [L], and K = 0.462 for foot-second units or 0.311 for meter-second units. Equation 7.15 is employed to calculate the pipe diameter after the design discharge, Q, and the sewer slope are known. The diameter determined by Eq 7.11 is so called *hydraulically required sewer diameter*. The calculated pipe size is often not commercially available. As a result, the next larger commercially available pipe size shall be used.

7.6.2. Arch (Elliptical) Sewer Hydraulics

Often the narrow clearance due to the exiting underground utilities sets the constraints to the sewer vertical profile. Between two manholes, a flat and wide sewer may fit into the narrowed corridor. As a result, an elliptical or arch pipe is selected as a replacement of the circular pipe. The flow through an elliptical or arch pipe in Figure 7.6 is dictated by the cross sectional geometry. For simplicity, an equivalent circular pipe may be used as an approximation in hydraulic computations.

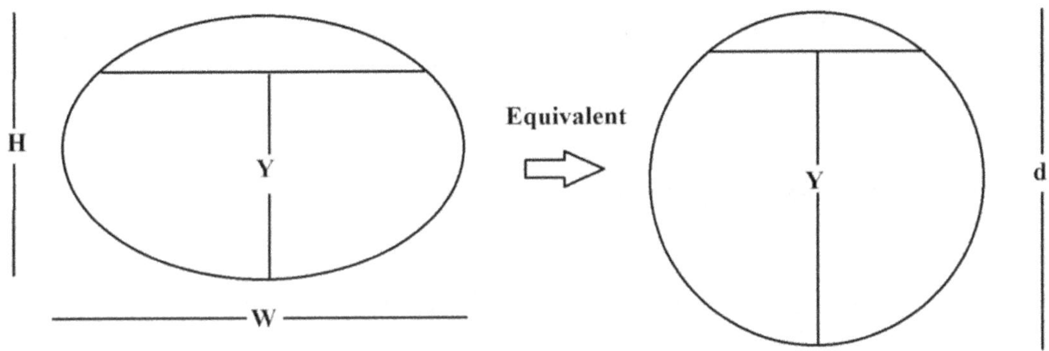

Figure 7.6. Elliptical Sewer.

The equivalent diameter is approximated by

$$d = 0.5\sqrt{HW} \tag{7.12}$$

in which H = rise of arch or elliptical sewer in [L] and W = span of arch or elliptical sewer in [L].

7.6.3. Box Sewer Hydraulics

As illustrated in Figure 7.7, the hydraulic parameters in a box sewer are related to the flow depth as:

$$A = BY \tag{7.13}$$

$$P = 2Y + B \tag{7.14}$$

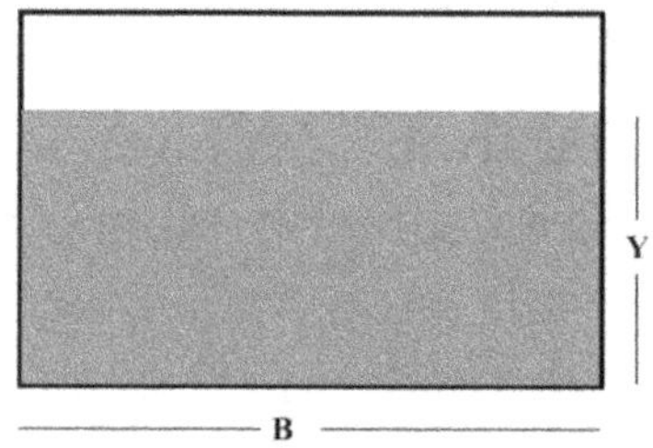

Figure 7.7. Box Sewer.

When a box sewer is selected, the width of the box sewer must be specified first. The hydraulic calculation is to provide the flow depth.

7.7. DESIGN CONSTRAINTS

As a trial and error process, the sewer system design is finalized by continuously adjusting sewer sizes and manhole drops until all the design criteria and constraints are satisfied, including:

(1) permissible flow velocity in sewer line between 3 and 20 ft/sec ,

(2) minimum earth coverage of 2 feet,

(3) minimum sewer diameter of 18 inches,

(4) minimum manhole drop of 0.20 foot

(5) minimum of 2 feet used for sewer trench bottom width

(6) earth side slope of 1V:1H used in sewer trench excavation,

(7) maximum manhole spacing of 400 feet, and

(8) the maximum ratio of normal flow depth to sewer diameter or height to be 0.8.

After all sewers are sized by open channel flow under the normal flow condition, the sewer system is further subject to a performance evaluation under the given tailwater at the system exist. The energy and hydraulic grade lines (EGL and HGL) will predict if any manholes in the sewer system are surcharged. Further modifications on the vertical profile may be required until the HGL for the design flow is kept below the ground at all manholes in the system

7.8. DESIGN PROCEDURES

Design of a sewer system begins with the placement of inlets and manholes. All sewer crowns between two manholes must be set at 2 foot or greater below the ground surface. Manhole drops can be introduced to reduce the water flow velocity or to avoid utility conflicts. Street inlets and storm sewers are is usually designed to pass the 2- to 5-year event. As shown in Figure 7.8, during a major event, the excess storm water would be carried by street gutters.

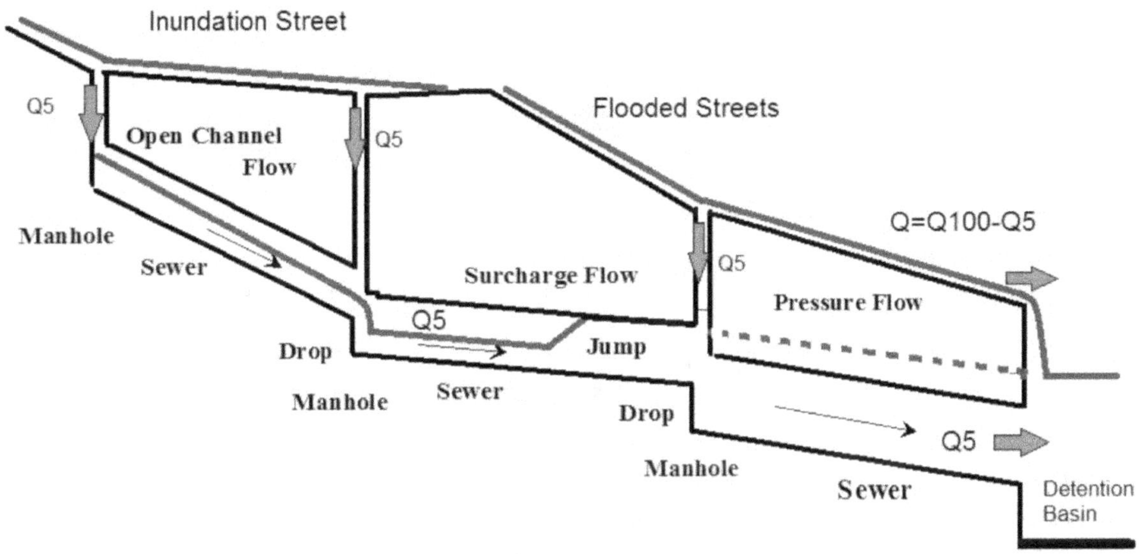

Figure 7.8. Dual Drainage Systems.

The sewer system collects the surface storm runoff flows from manholes. The sewer sizing process starts from the most upstream manhole and then marches downstream as the flow rate and the flow time are accumulated toward the system exit. At a manhole, the flow times along various flow paths must be first calculated, and then the longest one shall be considered as the design rainfall duration. The flow time in a sewer can be computed by either the flowing full velocity or the partially full flow velocity determined by Manning's formula. The difference in flow times between these two velocities is nearly negligible. For a flat or negatively sloped sewer, its flowing full velocity is used for hydraulic computations.

7.9. DESIGN DISCHARGE

An urban watershed is divided into small catchments based on the locations of the inlets. Often the street center lines along street crowns serve as catchment boundaries. The watershed in Figure 7.9 is divided into four catchments for four inlets.

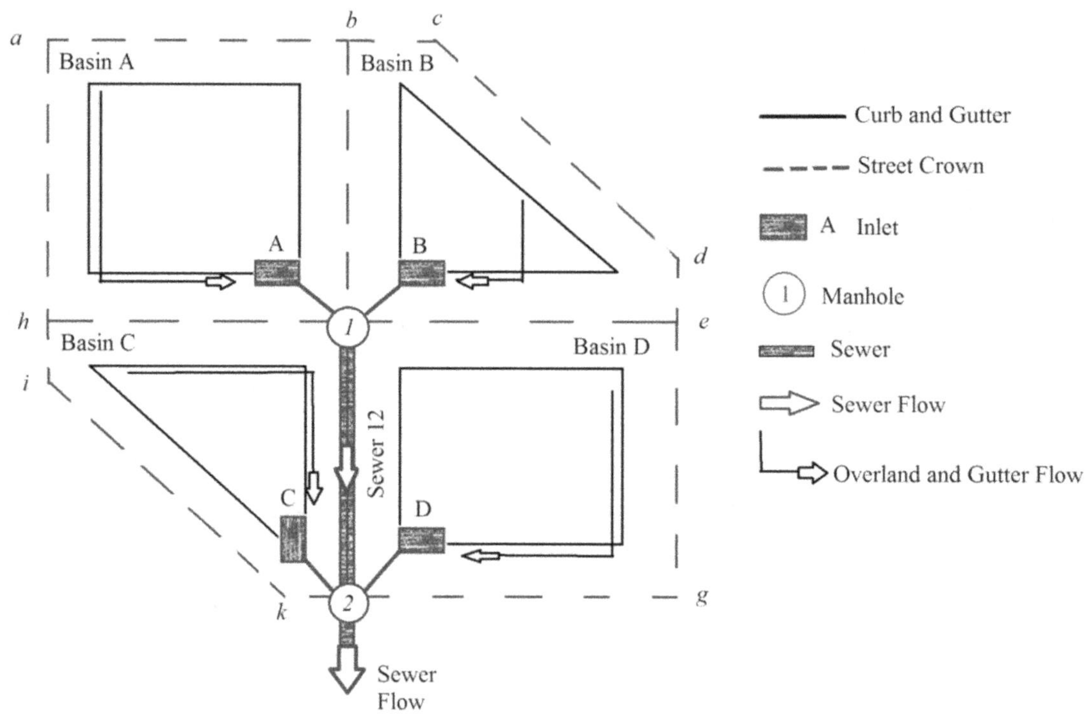

Figure 7.9. Design Flows for Inlets and Manhole.

The outgoing sewer from a manhole is sized based on the total design flow at the manhole. At Manhole 1, the tributary area is Area *acdeh in* Figure 7.9. At Manhole 2, the entire watershed is the tributary area. There are two flows coming into Manhole 1, therefore the longer time of concentration dictates the design flow at Manhole 1. For convenience, the Rational method is re-arranged using the effective area which is defined as the product of CA for each catchment.

$$A_e = \sum_{k=1}^{k=M} C_i A_1 \tag{7.15}$$

$$T_c = Max(T_{C1}, T_{C2}, T_{C3}, \dots T_{CM}) \tag{7.16}$$

$$I = \frac{K_1 P_1}{(K_2 + T_c)^{K_3}} \tag{7.17}$$

$$Q = IA_e \tag{7.18}$$

in which C= runoff coefficient, A= catchment area in [L^2], A_e = effective area in [L^2], T_c = time of concentration in [T], I = rainfall intensity in [L/T], Q= peak flow in [L^3/T], *i*= *i-th* catchment, k= counter in summation, M= number of catchments, and K_1, K_2, and K_3 = constants in rainfall intensity-duration-frequency (IDF) curve. Eq 7.4 is the design flow to size the outgoing sewer. The numerical routine from Eq's 7.15 to 7.18 is repeated at every manhole for sewer sizing.

EXAMPLE 7.1 The hydrologic parameters for Basins A and B are already given in Figure 7.10. Consider the rainfall IDF formula wih K_1P_1= 38.5, K_2=10, and K_3=0.789. Determine the peak flow at Manhole 1.

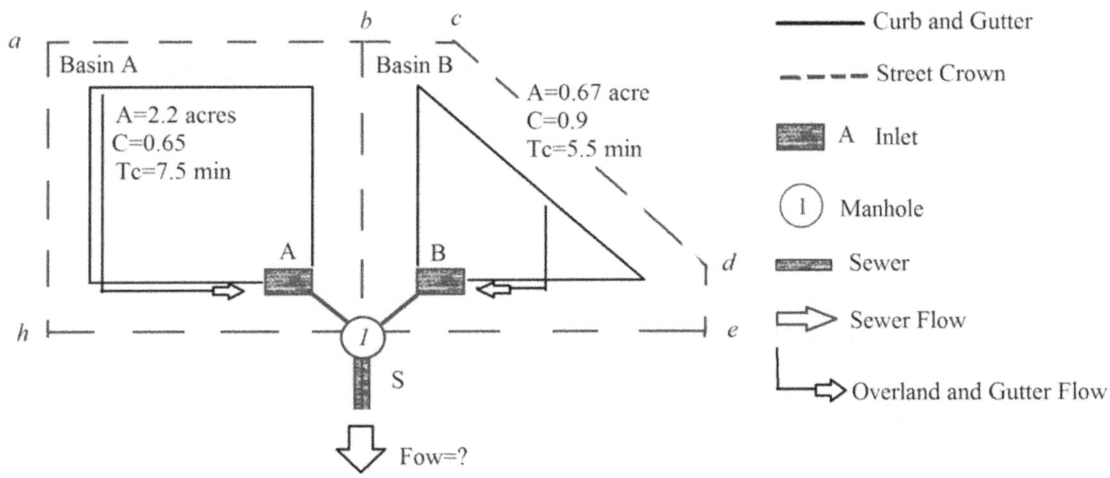

Figure 7.10. Design Flow at Manhole.

Solution:

$$A_e = \sum_{k=1}^{k=2} C_i A_1 = 0.65 \times 2.2 + 0.9 \times 0.67 = 2.03 \text{ acres}$$

$$T_c = Max(T_{C1}, T_{C2}, T_{C3}, ...T_{CM}) = \max(7.5, 5.5) = 7.5 \quad \text{min}$$

$$I = \frac{K_1 P_1}{(K_2 + T_c)^{K_3}} = \frac{38.5}{(10+7.5)^{0.789}} = 4.02 \text{ inch/hr}$$

$$Q = IA_e = 4.02 \times 2.03 = 8.2 \text{ cfs}$$

Having known the design flow of 8.2 cfs, we are ready to size the outgoing sewer.

EXAMPLE 7.2 Design a circular sewer to deliver a discharge of 8.2 cfs in EXAMPLE 7.1 on a slope of 1.0 % with Manning's coefficient of 0.012.

Solution:

1. Find the hydraulically required pipe size

$$d = (\frac{0.012 \times 8.2}{0.462\sqrt{0.01}})^{\frac{3}{8}} \times 12 = 16.5 \ inches$$

2. Use a 18-inch circular sewer that has a full flow capacity as

$$Q_f = \frac{1.486}{0.015} \times (1.5/4)^{\frac{2}{3}} \times (\frac{3.1416 \times 1.5^2}{4}) \times \sqrt{0.01} = 11.5 \ cfs$$

3. Determine the design flow condition in the 18-inch pipe.
The trial and error procedure begins with a guessed central angle. Aided with Eq's 7.9 to 7.12, we shall check if the central angle satisfies Manning's equation with the design flow of 8.2 cfs. For instance, Try θ=1.83 radians. Check Manning's equation for the design flow of 8.2 cfs as:

$$Y = \frac{d}{2}(1 - \cos\theta) = \frac{1.5}{2}(1 - \cos 1.83) = 0.94 \ ft$$

$$A = \frac{d^2}{4}(\theta - \sin\theta\cos\theta) = \frac{1.5^2}{4}(1.83 - \sin 1.83 \times \cos 1.83) = 1.17 \ ft^2$$

$$P = d\theta = 1.5 \times 1.83 = 2.75 \ ft$$

$$Q = VA = \frac{1.486}{0.012}(\frac{1.17}{2.75})^{2/3}\sqrt{0.01} \times 1.17 = 8.2 \ cfs \ \text{(Ok!)}$$

$$V = \frac{Q}{A} = \frac{8.2}{1.17} = 7.0 \ fps$$

$$T = d \sin\theta = 1.5 \times \sin 1.83 = 1.45 \ ft$$

$$Fr = \sqrt{\frac{Q^2 T}{gA^3}} = \sqrt{\frac{8.2^2 \times 1.45}{32.2 \times 1.17^3}} = 1.38 \text{ for normal flow.}$$

The above analysis is under the normal flow condition. In fact, a sewer pipe is also subject to downstream backwater effects. Under a surcharge condition, the sewer pipe likely becomes flowing full. Its full flow velocity is:

$$V_f = \frac{8.2}{3.1416 \times 1.5^2 / 4} = 4.64 \ fps$$

Sizing a sewer system starts from the most upstream manhole. Referring to Figure 7.6, Manhole 2 has three incoming flows, including two *local inlet flows* and one *sewer flow* from Manhole 1. For comparison, the flow times along these three flow paths must be calculated because the longest one is chosen for the design rainfall duration. For instance, from Manhole 1 to Manhole 2, the travel time is

$$T_s = L_s / (60 * V_s) \tag{7.19}$$

in which T_s= travel time through Sewer 12 in minutes, L_s = length for Sewer 12, and V_s = flowing full velocity through Sewer 12. The sewer flow velocity is assumed to be either full flow or partially full flow velocity. Often, the difference in flow times between these two velocities is numerically negligible. For this case, the time of concentration at Manhole 2 is

$$T_{M2} = max(T_{c3}, T_{c4}, T_{M1} + T_s) \tag{7.20}$$

in which T_{M2} = time of concentration at Manhole 2. T_{c3} = time of concentration from Basin 3, and T_{c4} = time of concentration from Basin 4. The design rainfall intensity, i_2, derived at Manhole 2 applies to Basins 1, 2, 3, and 4. Therefore, the design discharge at Manhole 2 is:

$$Q = i_2 (C_1 A_1 + C_2 A_2 + C_3 A_3 + C_4 A_4) \tag{7.21}$$

Repeat the above procedure to each manhole in the system until the outfall sewer pipe is sized. Having all sewers been individually sized, the sewer system is then ready for performance evaluation under the tailwater effects at the system outfall. The EGL and HGL will detect if any manholes and sewers are surcharged. To avoid reverse flows on the street from a surcharged manhole, it is necessary to continue modifying the sewer vertical profile until the HGL for the design flow is kept below the ground at all manholes in the system.

EXAMPLE 7.3 Continue EXAMPLE 7.2. Hydrologic parameters for Basins C and D are given in table 7.2. Determine the design discharge at Manhole 2

Table 7.2. Hydrologic Parameters for Basins C and D.

Catchment Parameters	Basin C	Basin D
Area (acres)	0.66	2.2
Tc to Basin Outlet (minutes)	5.5	7.5
Flow Time through Sewer AB	0	0
Runoff C	0.65	0.65

Sewer 12 has a length of 250 feet with a full flow velocity of 4.64 *fps* (see EXAMPLE 7.2). The travel time through Sewer 12 is:

$$T_s = \frac{L_s}{60V_s} = \frac{250}{4.64}\frac{1}{60} = 0.90 \text{ min}$$

$$T_{M2} = \max(T_{c3}, T_{c4}, T_{M1} + T_s) = \max(5.5,\ 7.5,\ 7.5 + 0.90) = 8.4 \text{ min}$$

$$I = \frac{K_1 P_1}{(K_2 + T_c)^{K_3}} = \frac{38.5}{(10 + 8.4)^{0.789}} = 3.87 \text{ inch/hr}$$

$$A_e = \sum_{k=1}^{k=M} C_i A_1 = \sum_{k=1}^{k=4} (0.65 \times 2.2 + 0.9 \times 0.66 + 0.65 \times 0.66 + 0.65 \times 2.2) = 3.89 \text{ acres}$$

$$Q = IA_e = 3.87 \times 3.89 = 15.03\ cfs$$

The outgoing sewer from Manhole 2 shall be sized to carry a peak flow of 15.03 cfs.

7.10. CASE STUDY

Sewer system A is located in the City of Denver, Colorado and illustrated in Figure 7.11. Use the following information to size the sewer pipes.

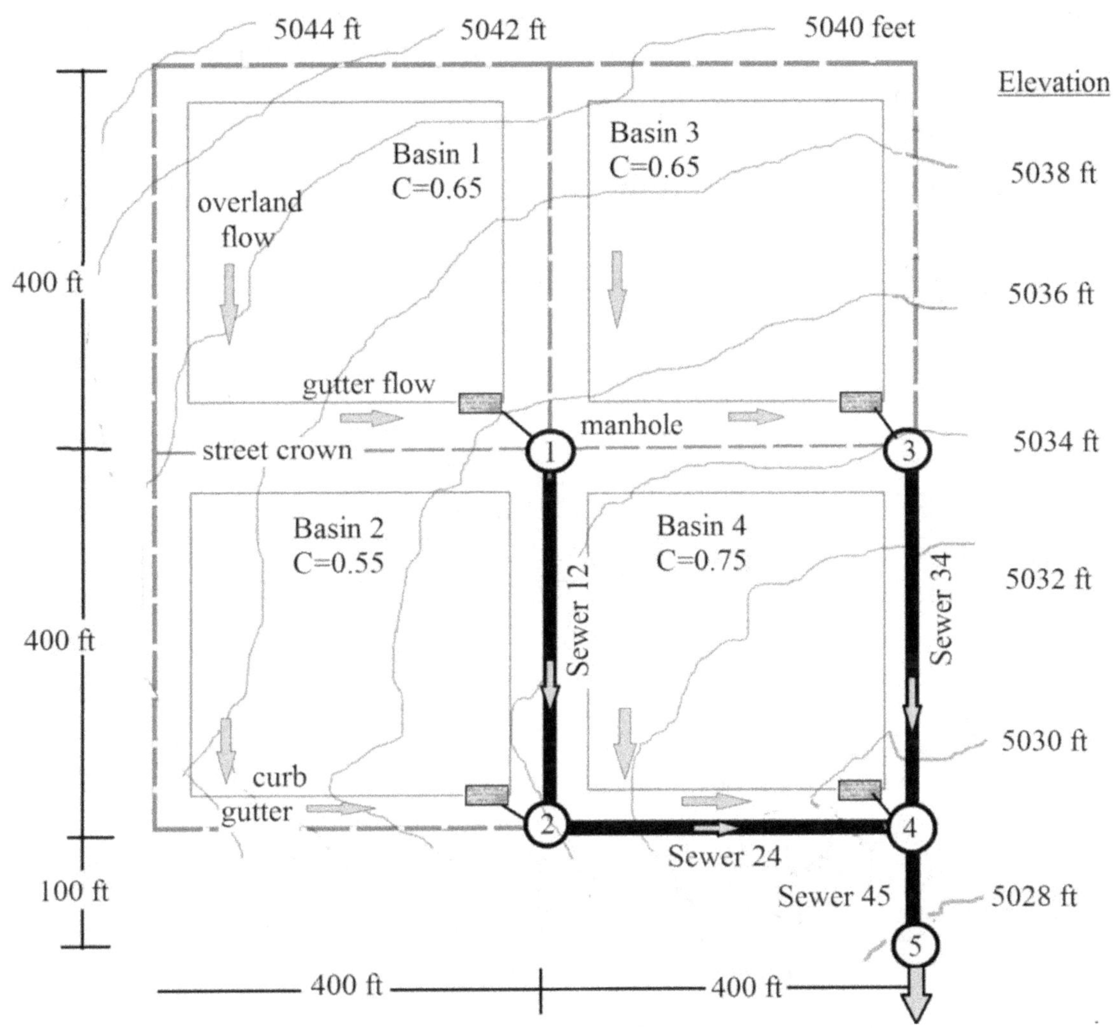

Figure 7.11. Layout of Storm Sewer System.

(a) The 5-yr design rainfall intensity, I (inch/hr), is given as:

$$I(inch/hr) = \frac{28.5P_1}{(10+T_d)^{0.789}}$$

in which P_1 = 1.35 inch and T_d=duration in minutes.

(b) The overland flow time, T_o, shall be calculated by the airport formula

$$T_o = \frac{0.393(1.1 - C)\sqrt{L_o}}{S_o^{\ 0.33}}$$

in which L_o= overland flow length ≤300 feet, C= runoff coefficient, S_o= overland flow slope in *ft/ft*.The swale and gutter flow velocities, V_f, can estimated by the SCS upland method with a conveyance coefficient of 20.

$V_f = 20\sqrt{S_f}$ in which V_f= flow velocity (*fps*) and S_f= ground slope in *ft/ft*.

(c) The tailwater surface at the system exit is at elevation 5022 feet.

Solution:

Basin ID	Area	Runoff Coef	Imp Ratio	Effective Area	Overland Flow			Swale Flov			Channel Flow			
					Slope	Length	Time	Length	SCS K	Time	Slope	Length	SCS K	Time
	acres			acres	%	ft	min	ft	fps	min	%	ft	fps	min
ID	A	C	Ia	CA	So	Lo	To	L2	K2	T2	S2	L3	K3	T3
1	3.67	0.65	0.70	2.39	1.00	300.00	14.03	100.00	20.00	0.83	1.00	400.00	20.00	3.33
2	3.67	0.55	0.60	2.02	0.50	300.00	21.55	100.00	20.00	1.18	1.00	400.00	20.00	3.33
3	3.67	0.65	0.70	2.39	1.00	300.00	14.03	100.00	20.00	0.83	0.50	400.00	20.00	4.71
4	3.67	0.75	0.80	2.75	0.50	300.00	13.72	100.00	20.00	1.18	1.50	400.00	20.00	2.72

Avg Slope	Exist Tc Comp	Future Tc Check	Storm Duration	Rainfall Intensity	Peak Flow
%	min	min	min	in/hr	cfs
Sa	TcComp	TcCheck	Td	I	Qp
1.00	18.20	12.12	12.12	1.37	3.28
0.75	26.07	15.41	15.41	1.19	2.41
0.75	19.58	13.42	13.42	1.30	3.10
1.00	17.62	10.32	10.32	1.49	4.12

2. Combined Peak Flows at Street Intersection

Street Gutter	Gutter Length	Gutter Slope	Local Catchment			Cumulative Parameters at Upst Design Pt.						Effective Area	Storm Duration
			Basin ID	Effective Area	Rainfall Duration	Incoming Gutter	Effective Area	Rainfall Duration	Travel Time Through Street				
									length	velocity	time		
	ft	ft/ft		acres	min		acre	min	ft	fps	min	acre	min
ID	Ls	Ss	ID	CA	Td	ID	CA	Td	Ls	Vs	Ts	CA	Td
12	**400**	**0.01**	1.0	2.39	12.12							**2.39**	**12.12**
24	**400**	**0.005**	2.0	2.02	15.41		2.02					2.02	15.41
					12.12	12	2.39	12.1	400.0	9.45	0.71	2.39	12.83
												4.41	**12.83**
34	**400**	**0.01**	3.0	2.39	15.41							**2.39**	**15.41**
45	**200**	**0.01**	4.0	2.75	13.42							2.75	15.41
						24	4.41	12.8	400.0	7.55	0.88	4.41	13.71
						34	2.39	15.4	400.0	9.07	0.74	2.39	16.15
												9.55	**16.15**

Rainfall Intensity	Peak Flow	Pipe Dia Req'd	Pipe Dia Used	Sewer Flow Velocity
in/hr	cfs	inch	inch	fps
I	Qp	D-req'd	D-used	Vs
3.34	**7.98**	**16.70**	**18.00**	**9.45**
3.26	**14.37**	**23.71**	**24.00**	**7.55**
3.00	**7.15**	**16.03**	**18.00**	**9.07**
2.93	**27.98**	**26.73**	**27.00**	**6.72**

7.11. HOMEWORK

Q 7.1 A sewer system shown in Figure HW 7.1 consists of three catchments and three manhole-sewer units and five catchments. Determine the design flows and sewer sizes. Design information is given as follows:

(a) Design Rainfall intensity formula

$$I(inch/hr) = \frac{40}{(10+T_d)^{0.76}}$$ in which T_d = duration in minutes and I= intensity in inch/hr.

(b) Design rainfall duration for estimating the peaking runoff from a catchment is the time of concentration of the catchment. The overland flow time shall be estimated by the airport formula, and the gutter flow velocity estimated by the SCS upland method for pavement surface.

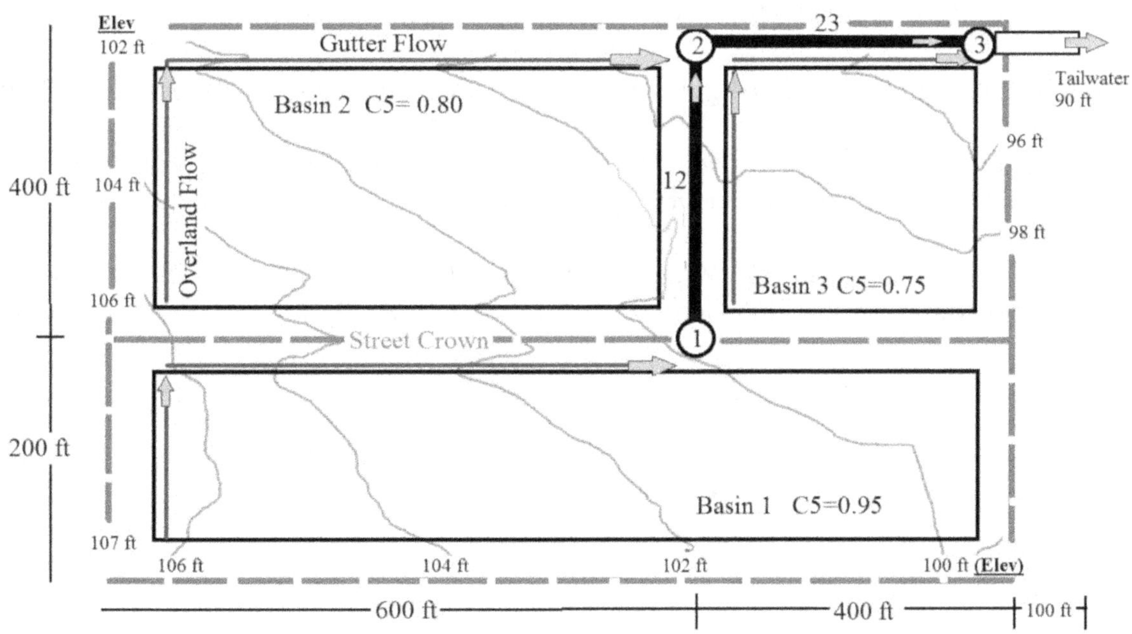

Figure HW 7.1. Sewer System Design for Home Work.

(c) Design constraints are:
(1) permissible flow velocity in a sewer between 20 feet/second and 3 feet/second ,
(2) minimum earth coverage of 2 feet
(3) minimum sewer diameter of 18 inches,
(4) minimum manhole drop of 0.50 foot

7.12. REFERENCES

American Iron and Steel Institute, (1980) "*Handbook of Steel Drainage and Highway Construction Product.*", Washington D.C.

American Iron and Steel Institute, (1980) "*Modern Sewer Design.*", Washington D.C.

American Society of Civil Engineers, (1979) "*Design and Construction of Sanitary and Storm Sewer.*" New York.

Guo, James C. Y. (1985) *"Technical Manual for UDSEWER Computer Model."* Research Report, Department of Civil Engineering, University of Colorado at Denver, Denver, Colorado.

Guo, James C.Y. (1989a) "*Auto Sizing Techniques for Storm Sewer System Design Using UDSWMM*", Proceedings for the National EPA Conference, Denver, Colorado.

Guo, James C.Y. (1989b). "*Energy Dissipation in Storm Sewer System*", Proceedings of ASCE Conference on Stormwater Modeling and Management, held at Denver, Colorado in May.

Rossman, L. (2005), EPA SWMM5 User Manual, http://www2.epa.gov/

SWMM5 (2005), http://www2.epa.gov/water-research/storm-water-management-model-swmm

UDFCD Manual (2010), "*Storm Water Drainage Design Criteria Manual*", Vol 1 and Vol 2, Urban Drainage and Flood Control District Denver, Colorado.

Yen, B.C., Wenzel, H.G., Mays, L.W., Tang, W.H. (1976) *"New Models for Optimal Sewer System Design."* Proceedings for the EPA Conference on Environmental Modeling and Simulation, Cincinnati, Ohio, July.

CHAPTER 8
EVALUATION AND COST FOR SEWER SYSTEM

Each individual sewer pipe is sized by its normal flow depth under the assumption of no backwater effects at all. Having all sewer pipes sized in a system, the engineer must calculate the energy and hydraulic grade lines (EGL and HGL) through the system to identify if any manholes are surcharged to the ground elevation. The water surface profile computation is performed starting from the system outfall to the most upstream manhole. Various tailwater conditions at the system outfall shall be evaluated to quantify the impact on the sewer line's performance. The water surface profile through a sewer line represents its *hydraulic gradient line (HGL)*. The difference between the HGL and its *energy gradient line (EGL)* is the flow kinetic energy, or velocity head. A proposed sewer line becomes acceptable if the design flow condition at all manholes and through all sewers satisfy the design criteria.

The construction cost for a proposed sewer line depends on the materials, earth volume excavation, labor and management. In the Chapter, a detailed discussion is given on the earth volume estimates along a sewer trench.

(a) HGL above Ground | (b) Surcharged Manhole

Figure 8.1. Surcharged Sewer System.

8.1. SEWER-MANHOLE ELEMENT

To analyze the EGL and HGL in a sewer system, it is necessary to recognize that the basic element in a sewer system is a manhole-sewer unit, i.e. a *manhole connected to an outgoing sewer*. As shown in Figure 8.2, such a *manhole-sewer element* is repeated to complete the sewer system. A manhole-sewer element has four distinct sections. They are Sections 1, 2, 3, and 4.

Section 1 is the tailwater located immediately downstream of the sewer exit.

Section 2 is located immediately upstream of the sewer exit.

Section 3 is located immediately downstream of the sewer entrance; and

Section 4 is located immediately upstream of the sewer entrance.

When applying the energy principle from Sections 1 to 2, the exit loss must be counted. Flowing through Sections 2 to 3, the friction loss must be included. From Sections 3 to 4, the entrance losses must be added. From Sections 4 to 1, the water flow is subject to the manhole losses that are dictated by the manhole configurations.

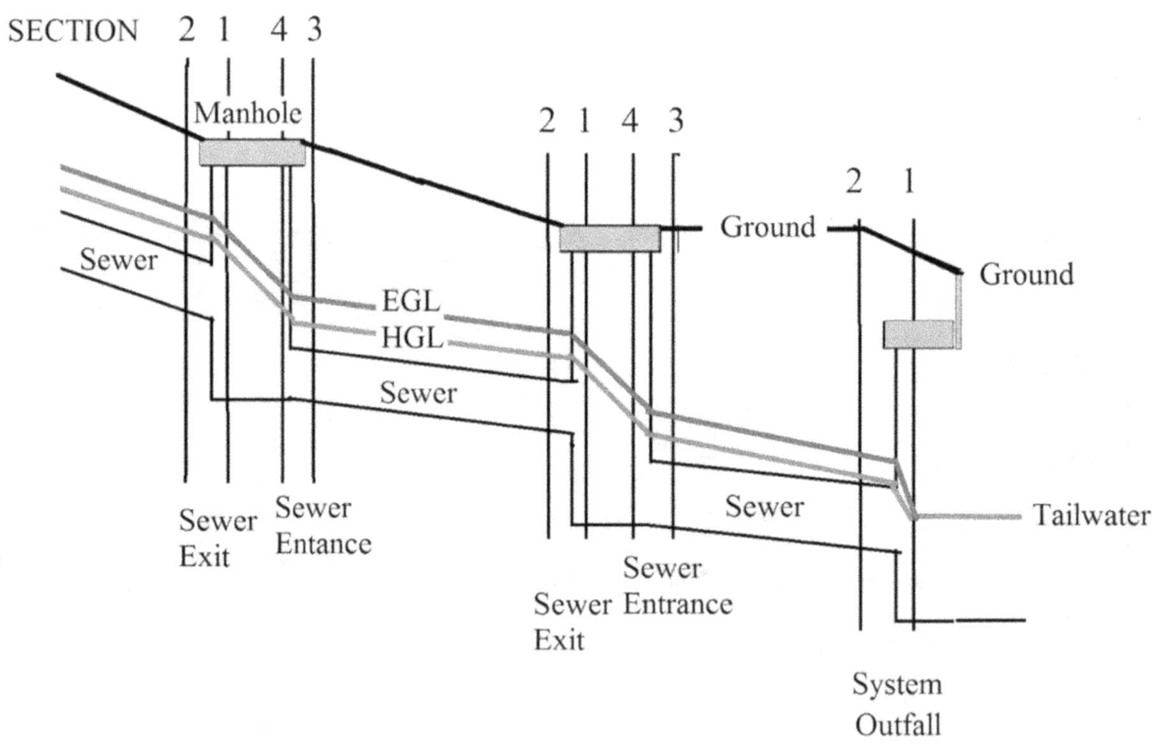

Figure 8.2. Manhole-Sewer Element in a Sewer System.

As shown in Figure 8.2, the computation of EGL starts from the sewer outfall. The system outfall is the most downstream Section 1 where the tailwater elevation must be known. In this Chapter, the energy balance in a manhole-sewer element is simplified using the full flow condition throughout the sewer pipe.

The analysis of energy balance divided into two steps:

(1) *Sewer Hydraulics:* Energy losses through the sewer pipe include:

(a) exit loss from Section 1 to Section 2,

(b) friction loss from Section 2 to Section 3, and

(c) entrance loss from Section 3 to Section 4.

(2) *Manhole (Juncture) Hydraulics*: Energy losses through the manhole, from Section 4 to Section 1, include:

(a) lateral loss due to incoming lateral, and

(b) bend loss due to the change in flow direction.

After the EGL at Section 1 becomes known, repeat Steps (1) and (2) for the next manhole-sewer element.

8.2. SEWER HYDRAULICS

8.2.1. Energy Analysis from Section 1 to Section 2

With the known tailwater condition, EGL= E_1 at Section 1 (i.e. water surface elevation in the downstream water body). As shown in Figure 8.2, between Sections 1 and 2, the energy balance is written as:

$$E_{12} = EGL_1 + K_x \frac{V_f^2}{2g} \tag{8.1}$$

in which E_{12} = energy at Section 2 determined from the given tailwater at the outfall, V_f = full flow velocity at sewer exit, K_x = exit energy loss coefficient ranging from 0.5 to 1.0. As illustrated in Figure 8.3, the EGL at Section 2 is not solely determined by the tailwater condition as E_{12}, because the manhole drop at Section 2 may elevate the sewer exit. A significant drop can introduce a discontinuity to the flow line. It implies that the new EGL may replace E_{12}. Based on the vertical profile, the HGL at Section 2 is:

$$E_e = Z_e + \frac{V_f^2}{2g} + D \tag{8.2}$$

Where E_e = required EGL at sewer exit based on the vertical profile, Z_e= invert elevation in [L] at sewer exit in [L], and V_f= full flow velocity, and D= water depth in [L] at the sewer exit. Equation 8.1 presents a tailwater EGL while Eq 8.2 represents a headwater EGL. As a result, the EGL at Section 2 is:

$$EGL_2 = Max\ (E_{12}, E_e) \tag{8.3}$$

If E_{12} is selected, then the EGL is on a continuous grade. If E_e is selected, the manhole drop has created a discontinuity on the EGL. The corresponding HGL or the water surface elevation at Section 2 is calculated as:

$$HGL_2 = EGL_2 - \frac{V_f^{\ 2}}{2g} \tag{8.4}$$

Where EGL_2 = elevation of EGL in [L] at Section 2, and HGL_2 = water surface elevation in [L] at Section 2.

EXAMPLE 8.1 A 24-inch circular sewer in Figure 8.3 carries a discharge of 24 cfs. The downstream tailwater EGL is: EGL_1 = 90.45 feet. Let K_x=0.5. Evaluate two cases at the incoming sewer exit: (a) invert elevation at 96 feet, and (b) invert elevation at 100 feet. Under the flowing full condition, estimate EGL_2 at the sewer exit for both cases.

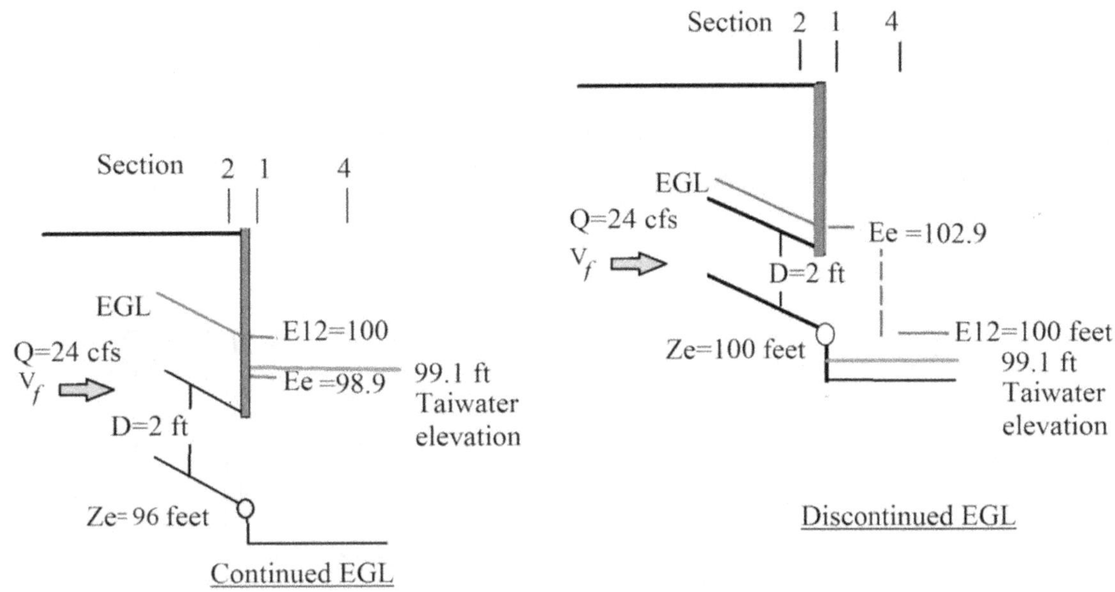

Figure 8.3. Tailwater and Flow Conditions at Sewer Exit.

<u>*Solution:*</u> Under the flowing full condition, the flow velocity is calculated as:

$$V_f = \frac{Q}{A} = \frac{24.0}{3.14 \times 2.0^2 / 4.0} = 7.64 \; fps$$

$$E_{12} = 99.1 + \frac{7.64^2}{2.0 \times 32.2} = 100 \, feet$$

When Z_e= 96 feet, the flow condition at the exit is

$$E_e = Z_e + D + \frac{V_f^{\,2}}{2g} = 96.0 + 2.0 + \frac{7.64^2}{2.0 \times 32.2} = 98.9 \; feet$$

As a result, the EGL at Section 2 is dominated by the EGL at Section 1 as:

$$EGL_2 = Max(E_{12}, E_e) = Max(100.0, 98.9) = 100.0 \, feet$$

The corresponding water surface at the exit is

$$HGL_2 = 100.0 - \frac{7.64^2}{2.0 * 32.2} = 99.1 \; feet$$

8.2.2. Energy Analysis from Section 2 to Section 3.

From Sections 2 to 3 in Figure 8.4, the energy balance includes the friction loss that can be calculated by Manning's equation under the full flow condition.

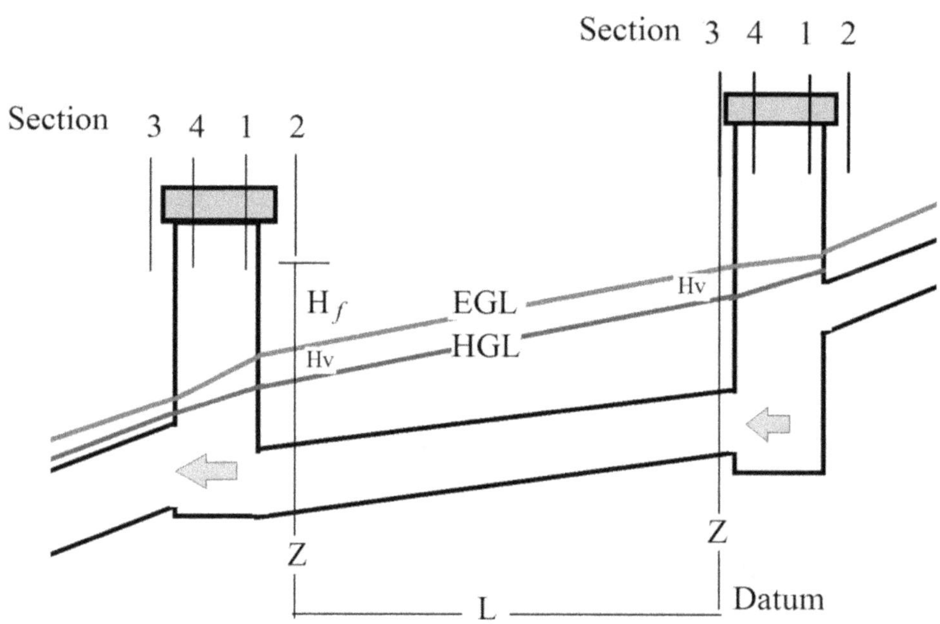

Figure 8.4. Energy Grade Line between Sections 2 to 3.

The energy principle between Sections 2 and 3 is written as:

$$HGL_3 + H_v = HGL_2 + H_v + H_f \tag{8.5}$$

$$H_v = \frac{V_f^2}{2g} \tag{8.6}$$

$$H_f = S_f L \tag{8.7}$$

$$S_f = \frac{n^2 V_f^{\ 2}}{2.22(D/4)^{4/3}} \tag{8.8}$$

$$HGL_3 = HGL_2 + H_f \tag{8.9}$$

in which V_f= full flow velocity in [L/T], H_v = kinetic energy in [L], Z= invert elevation in [L], H_f = friction losses in [L], S_f = friction slope in [L/L], L= length of sewer pipe in [L], and the subscript "2" represents variables associated with Section 2 etc.

EXAMPLE 8.2 Continue EXAMPLE 8.1. The incoming sewer of 24-inch in diameter carries a discharge of 24 cfs. The downstream tailwater EGL is E_1 = 102.9 ft. Plot the HGL across the concrete sewer line of 200 ft long on a slope of 1%.

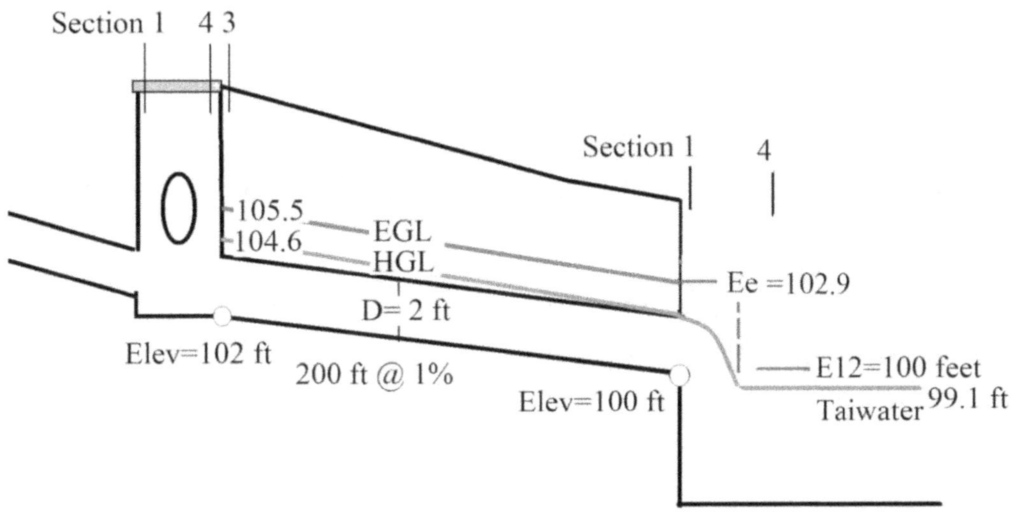

Figure 8.5. Tailwater and Flow Conditions at Sewer Exit.

Solution: Under the flowing full condition, the balance of EGL is calculated as:

$$V_f = \frac{Q}{A} = \frac{24.0}{3.14 \times 2.0^2 / 4.0} = 7.64\ fps$$

HGL₂=102 ft and HGL₂= 102.9 ft.

$$S_f = \frac{n^2 V_f^{\ 2}}{2.22(D/4)^{4/3}} = \frac{0.014^2 \times 7.64^2}{2.22(2/4)^{4/3}} = 0.013$$

$$H_f = S_f L = 0.013 \times 200 = 2.6\ ft$$

$$HGL_3 = HGL_2 + H_f = 102 + 2.6 = 104.6\ ft$$

$$EGL_3 = HGL_3 + \frac{V_f^2}{2g} = 104.6 + \frac{7.64^2}{2\times 64.4} = 105.5\ ft$$

8.2.3. HGL Analysis from Sections 3 to 4

The HGL analysis from Sections 3 to 4 in a sewer-manhole element essentially deals with the entrance loss as:

$$EGL_4 = EGL_3 + K_e \frac{V_f^{\ 2}}{2g} \tag{8.10}$$

in which K_e= entrance loss coefficient between 0.25 to 0.5, depending on the configurations of the entrance

EXAMPLE 8.3 Continue EXAMPLE 8.2. EGL_3=105.5 feet at Section 3. Considering K_e=0.2, determine the EGL at Section 4.

$$EGL_4 = EGL_3 + K_e \frac{V_f^{\ 2}}{2g} = 105.5 + 0.2 \times \frac{7.64^2}{2\times 32.2} = 105.68\ feet$$

8.3. MANHOLE HYDRAULICS

A manhole may have several incoming laterals, but only one outgoing sewer. Referring to Figure 8.2, the EGL analysis across a manhole from, Sections 4 to 1, starts with the known energy, EGL_4. There are two types of juncture losses at a manhole. They are:

(1) *bend losses* due to the change in the flow direction
(2) *lateral losses* caused by incoming lateral sewers on the trunk line only.

At a manhole, the incoming sewer is always subject to a bend loss, depending on the angle between the sewer lines. If there is incoming laterals at the manhole, only the trunk line is incurred with lateral losses as:

$$EGL_1 = EGL_4 + H_b + H_m \quad \text{(bend and lateral losses)} \tag{8.11}$$

$$EGL_1 = EGL_4 + H_b \quad \text{(bend loss only)} \tag{8.12}$$

in which HGL_1 = elevation of HGL at Section 1. Bend losses, H_b, is often estimated as a fraction of the full flow velocity head in the incoming sewer. Bend loss equation states:

$$H_b = K_b \frac{V_f^2}{2g} \tag{8.13}$$

in which V_f = the incoming full flow velocity in the incoming sewer into the manhole. The value of K_b is determined by the angle between the incoming flow direction and the outgoing flow direction at the manhole. Table 8.1 lists bend loss coefficients. It is noticed that a straight-through alignment has a bend loss coefficient of 0.05.

Lateral losses are only applicable to the trunk (main) line sewers. Lateral losses count for the additional turbulence caused by the branch sewers. The value of lateral loss coefficient, K_m, in Table 8.1 is determined by the angle between the branch sewer line and the main line. Lateral losses, H_m, are estimated by:

$$H_m = \frac{V_{fo}^2}{2g} - K_m \frac{V_{fi}^2}{2g} \tag{8.14}$$

in which V_{fo}= full flow velocity of outgoing sewer at manhole, and V_{fi} = full flow velocity of incoming lateral at manhole.

As illustrated in Figure 8.6, the manhole in Case one has only one incoming lateral. The bend loss is estimated by a coefficient of 0.28 determined by the angle of 45-degree. Case two involves three incoming sewers. Each incoming sewer has its own bend loss coefficient due to its own angle relative to the outgoing sewer. But, the one along the main line has an additional loss due to the two laterals. According to the angles of the two laterals, we have two different lateral loss coefficients. To be conservative, the smaller one is used because it results in a higher loss according to Eq 8.14.

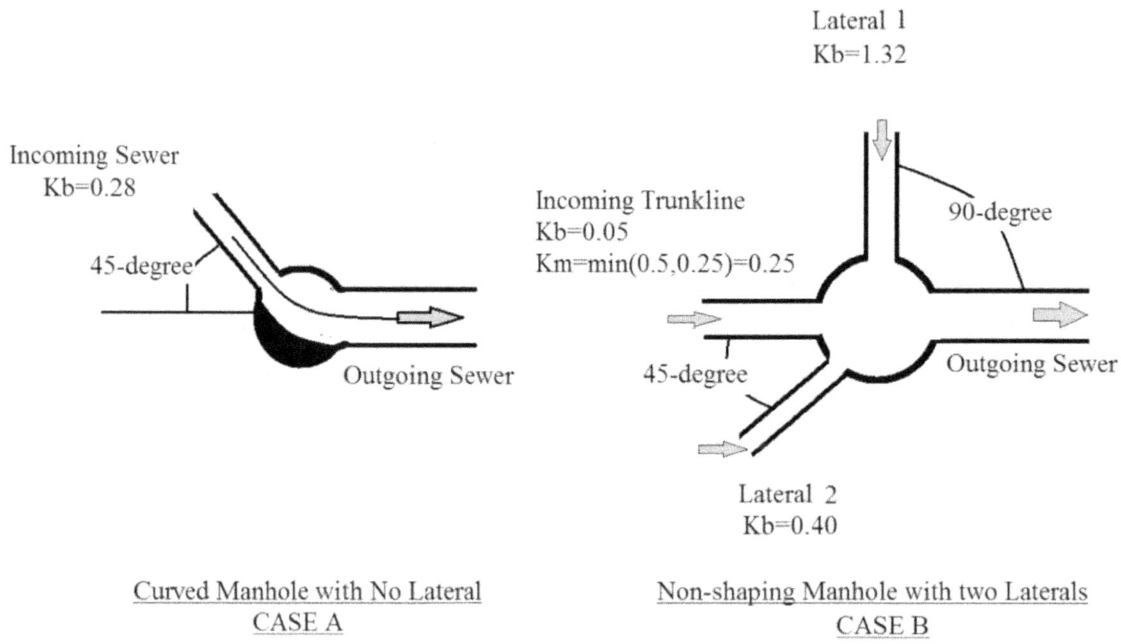

Figure 8.6. Lateral Losses and Bend Losses at a Manhole.

Table 8.1. Loss Coefficients at Manholes.

Angle in Degree	Bend Loss Coefficient for Curved Deflector in the Manhole	Bend Loss Coefficient for Non-shaping Manhole	Lateral Loss Coefficient on Main Line Sewer
Straight Through	0.050	0.050	Not Applicable
22.500	0.080	0.100	0.750
45.000	0.280	0.400	0.500
60.000	0.460	0.640	0.350
90.000	1.000	1.320	0.250

EXAMPLE 8.4 Continue EXAMPLE 8.3. The diameter of the incoming sewer line is an 18-inch circular pipe with a full flow velocity of 12.6 *ft/second*. The incoming angle is 45 degrees. The lateral is tied into the trunk line with an angle of 90 degrees with a full flow velocity of 9.5 *ft/second.* Knowing that EGL_4= 105.68 feet. Determine the juncture loss.

Solution: As illustrated in Figure 8.7, the bend and juncture losses are:

$$H_b = K_b \frac{V_f^2}{2g} = 0.4 \times \frac{12.6^2}{2 \times 32.2} = 0.98\ ft$$

$$H_m = \frac{V_{fo}^{\ 2}}{2g} - K_m \frac{V_{fi}^{\ 2}}{2g} = \frac{7.64^2}{2 \times 32.2} - 0.25 \times \frac{9.5^2}{2 \times 32.2} = 0.56 ft$$

$$EGL_1 = 105.68 + 0.98 + 0.56 = 107.22\ feet$$

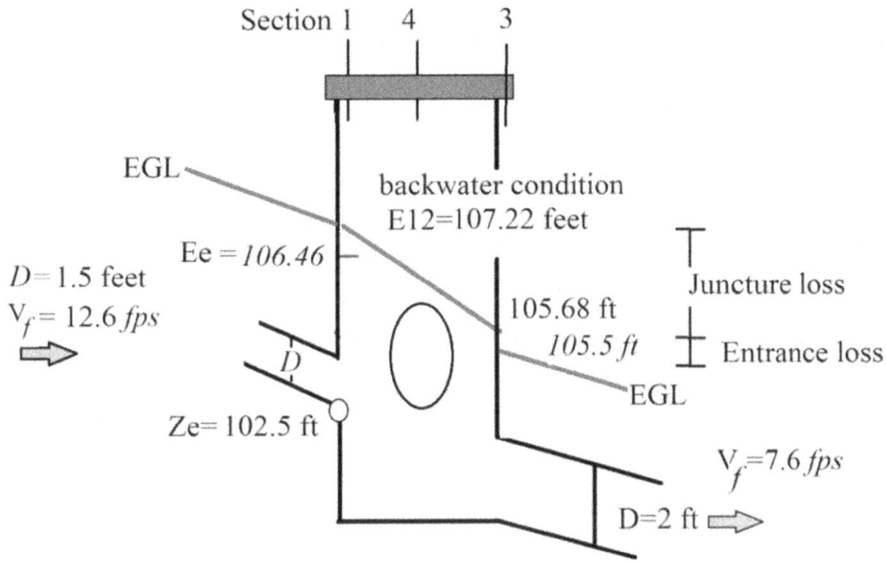

Figure 8.7. Juncture Losses.

EXAMPLE 8.5 Continue EXAMPLE 8.4. Referring to Figure 8.7, the backwater condition at Section 4 is HGL_4= 107.22 ft. Determine the EGL_1 at Section 1.

Solution: Z_e= 102.5 feet, D= 1.5 feet, and V_f= 12.6 *fps*

$$E_{12} = 107.22 feet$$

$$E_e = Z_e + D + \frac{V_f^{\ 2}}{2g} = 102.5 + 1.5 + \frac{12.6^2}{2.0 \times 32.2} = 106.46\ feet$$

As a result, the EGL at Section 2 is dominated by the EGL at Section 1 as:

$$EGL_2 = Max(E_{12}, E_e) = Max(107.22, 106.46) = 107.22\ feet$$

$$HGL_2 = 107.22 - \frac{12.6^2}{2.0 \times 32.2} = 104.75 \ feet$$

Having known EGL_2, the computational procedure is repeated from Sections 2 to 3, and then Sections 3 to 4 etc..

8.4. SEWER TRENCH

The wall thickness formula for ASTM C76 Reinforced Concrete Circular Sewers states:

$$T\ (inch) = \frac{D\ (in\ inch)}{12} + 1 \tag{8.15}$$

where T= wall thickness in inches, and D= inner diameter in inches. Sewer trench in Figure 8.8 must be sized to work with machines and tools. The bottom width of a sewer trench is set to be the outer diameter of the sewer pipe plus two (2) feet if the sewer is smaller than 48 inches in diameter, or four (4) feet if the sewer is greater than or equal to 48 inches. As illustrated in Figure 8.8, the height of manhole tube is calculated from the ground elevation to the lowest invert elevation among incoming and outgoing sewers at the manhole.

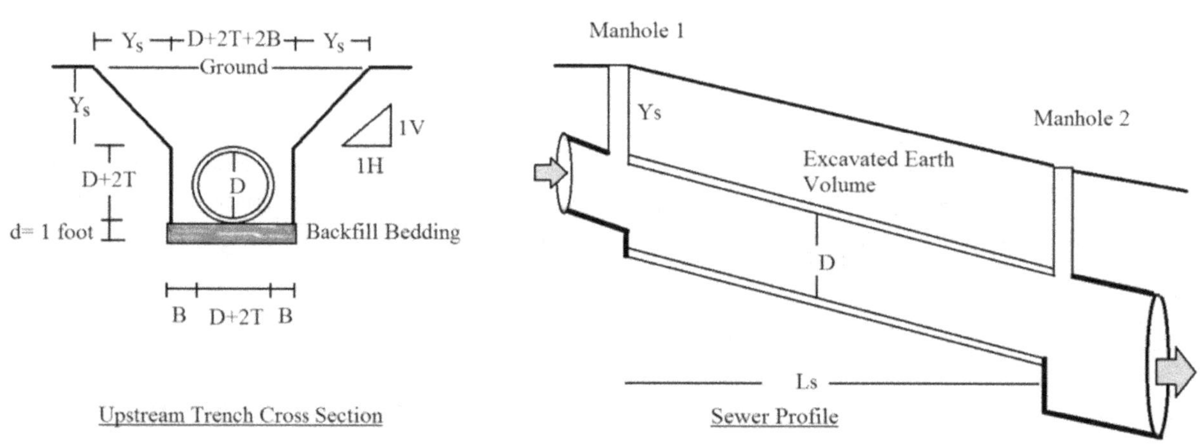

Figure 8.8. Excavation of Sewer Trench.

The excavated earth volume of a sewer trench is estimated by the average trench cross sectional area times the length of the sewer. For a noncircular pipe, an equivalent pipe diameter is used. The engineer can further estimate the cost when the unit prices for sewers, manholes, and earth excavation are available.

EXAMPLE 8.6 As shown in Figure 8.9, both upstream and downstream sections have a soil cover of 2 feet. Determine the excavated soil volume.

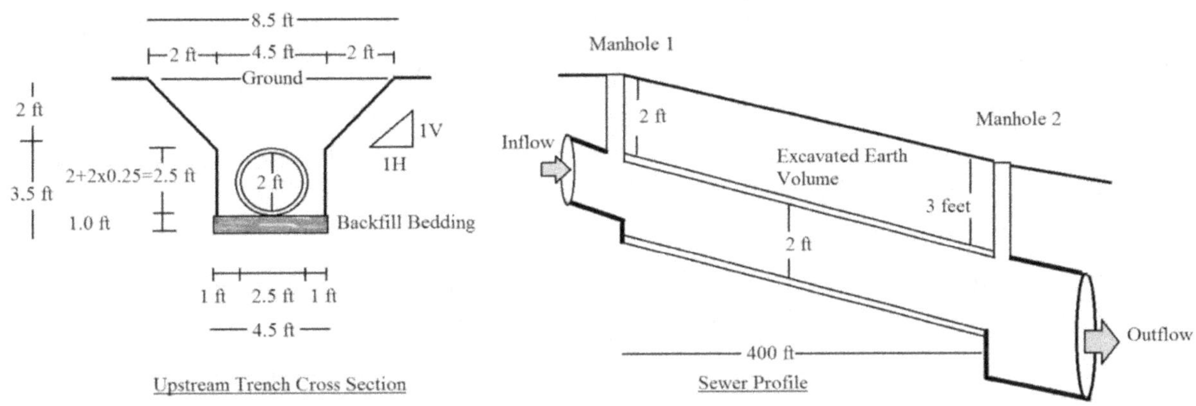

Solution:

$$B = 1 \text{ foot because } D < 48 \text{ inches}$$

$$T = \frac{24}{12} + 1 = 3 \text{ inches} = 0.25 \text{ foot}$$

$$D + 2T + 2B = 2 + 2 \times 1 + 2 \times 0.25 = 4.5 \text{ feet}$$

The upstream and downstream cross sectional areas are calculated as:

$$A_u = 4.5 \times 3.5 + \frac{8.5 + 4.5}{2} \times 2 = 28.75 \text{ feet}^2 \text{ (upstream section)}$$

$$A_u = 4.5 \times 3.5 + \frac{10.5 + 4.5}{2} \times 3 = 38.25 \text{ feet}^2 \text{(downstream section)}$$

The excavated earth volume is calculated with the average cross sectional area and length of trench as:

$$A_u = \frac{28.75 + 38.25}{2} \times 400 = 13{,}400 \text{ feet}^3$$

8.5. CLOSING

Sewers are sized individually, and then evaluated as a system subject to the downstream tailwater condition. Hardly can the flow in a sewer be uniform because the downstream and upstream disturbances to the flow. In this Chapter, the simplified approach was derived to calculate the HGL along a sewer line under the assumption of full flow in every sewer. This assumption is well justified when the sewer system is under backwater effects. A high tailwater depth at the outfall point can result in surcharged flows through the system. On top, rough pipe walls and manholes add more energy losses to the system. The computer model, *UDSEWER*, supported by the Urban Drainage and Flood Control District at Denver, Colorado (WWW.UDFCD.Org) provides flow estimates using the Rational method, sewer sizing using the normal flow depth, and calculations of HGL and EGL based on the detailed flow analyses, including open channel flow, surcharged flow, and flowing full flow along sewer lines.

On top of hydrology and hydraulic calculations, UDSEWER summarizes the lengths, widths, diameters, and earth volume along a sewer system for a cost estimate.

8.6. HOMEWORK

Q8.1 A sewer system is built as shown in Figure HW 8.1. All sewers are circular concrete pipes. The tailwater elevation is 105 feet. Plot HGL and EGL for the given diameter and design flow.

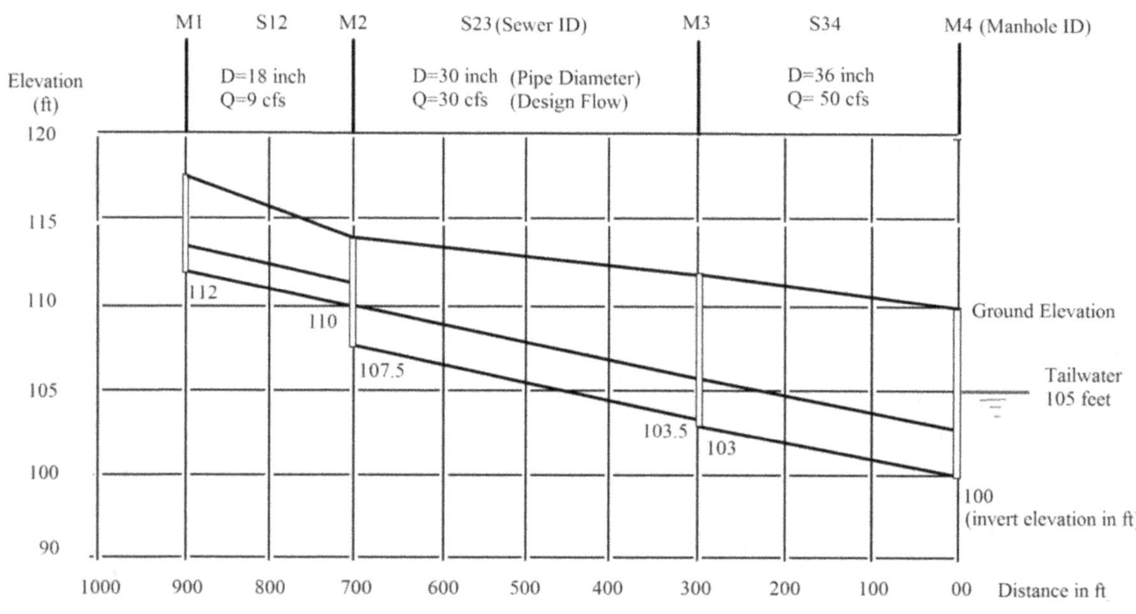

Figure HW 8.1. HGL and EGL for Sewer System.

Q8.2 Referring to Figure HW 8.1, determine the soil excavation volume in cubic yards.

8.7. REFERENCES

American Iron and Steel Institute, (1980) "*Modern Sewer Design.*", Washington D.C.

Guo, James C. Y. (1985) *"Technical Manual for UDSEWER Computer Model."* Research Report, Department of Civil Engineering, University of Colorado at Denver, Denver, Colorado.

Guo, James C.Y. (1989b). "*Energy Dissipation in Storm Sewer System*", Proceedings of ASCE Conference on Stormwater Modeling and Management, held at Denver, Colorado in May.

Rossman, L. (2005), EPA SWMM5 User Manual, http://www2.epa.gov/

SWMM5 (2005), http://www2.epa.gov/water-research/storm-water-management-model-swmm

UDFCD Manual (2010), "*Storm Water Drainage Design Criteria Manual*", Vol 1 and Vol 2, Urban Drainage and Flood Control District Denver, Colorado.

CHAPTER 9
INTRO TO COMPUTER MODEL: UDSEWER

9.1. INTRODUCTION

The computer model, UDSEWER, was developed in 1994 by the University of Colorado Denver. Over the years, UDSEWER has been supported by the Urban Drainage and Flood Control District (UDFCD). Since 1994, UDSEWER has been continuously enhanced with its friendliness and capability. UDSEWER 2009 Version is the latest and can be downloaded from the website: www.UDFCD.org as the recommended design tool for sizing sewers and analyzing flows in a drainage system.

UDSEWER applies the Rational method to size a sewer system and then tabulates the excavated earth volumes, pipe diameters, and manhole heights for cost estimates. The input parameters include: *(1) rainfall intensity-duration formula or table, (2) design constraints, (3) design flow information at manholes, and (4) sewer parameters*. At a manhole, the design flow can be calculated by the *Rational method*, or specified by the user. Sewers can be new or existing with a known diameter. New sewers are sized using circular pipes and the existing sewers can be circular, box, or elliptic (arch) pipes. The slopes of sewers can be positive, negative, or flat. UDSEWER will also calculate the energy and hydraulic grade lines by open channel flows, pressure flows, and surcharge flows. For a large watershed, the user may divide the watershed into two smaller sub-catchments. The Option of "Known Design Flow" will transfer an off-site flow into the UDSEWER model for a continuous simulation using the Rational method to accumulate flows and flow times. UDSEWER generates text output files and graphic plots of sewer vertical profiles with EGL and HGL.

9.2. GETTING STARTED (INSTALLATION)

To download the UDSEWER package, visit the website: http://udfcd.org/software.. Click on the decompressed package to download: UD-Sewer 2009 | v1.4.0 May-2014 (zip, 6.23 MB). The package includes the software and user manual.

Save the file into a local directory. Next, unzip it. Run the package to install the working file and example files. OF course, the user can always assign a different directory for file management.

9.3. HOW TO START UDSEWER

UDSEWER provides a graphic utility input (GUI) for input, output, and program operations. To access the GUI as shown in Figure 9.1, click on the *program icon* to enter the page in Figure 9.1.

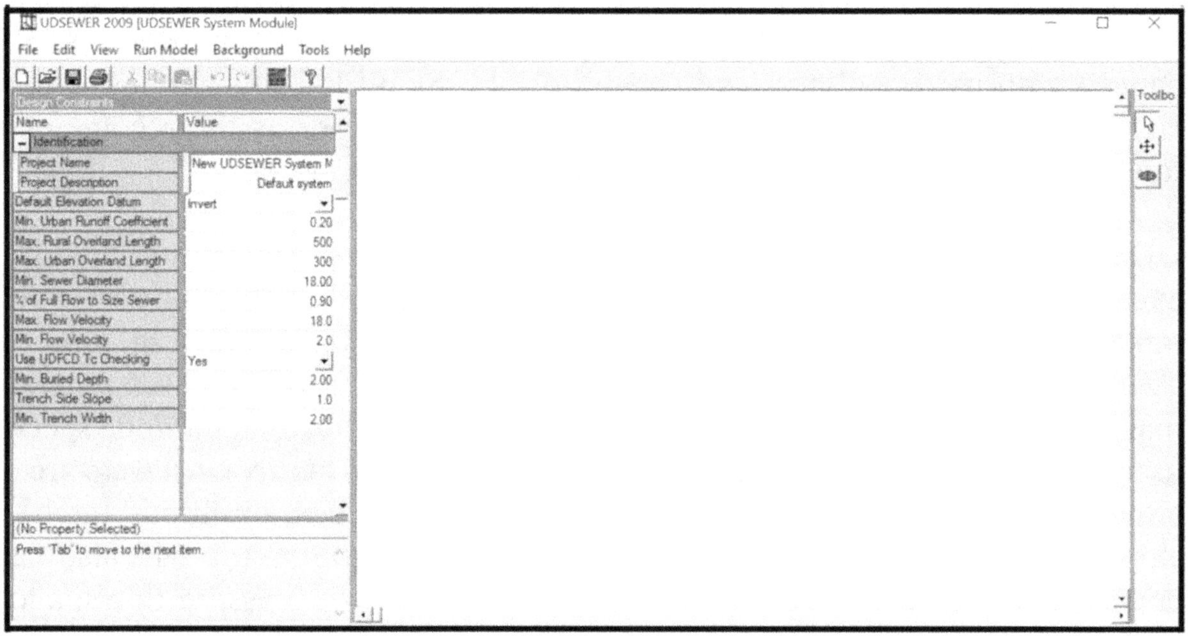

Figure 9.1. Graphic Utility Input (GUI).

9.4. INPUTS TO UDSEWER

Each input file represents one sewer system. Three sets of input data shall be provided by the user, including (1) *General Inputs* for design rainfall, design constraints, and default values, (2) *Sewer-Manhole Inputs* for each sewer-manhole unit, and (3) *Tailwater Input* at the system outfall. Output files include tabulations of sewer-manhole input parameters and hydraulic performance, and graphic plots of sewer vertical profile with EGL and HGL.

9.4.1. Setting the Global Properties

This package provides on-screen data file editing and creating. Data entry includes (1) *Design Constraints (data input), (2)Manhole-Sewer Layout (graphic input), (3)Manhole and Sewer Design Information (data input),* including design flows, tributary watershed hydrology, and sewer parameters.

From "*File Menu*", you can access "*Design Constraints*" as illustrated in Figure 9.2. "Click on "*Design Constraints*" to switch to "*Rainfall Constraints*" or a couple of user's preference on numerical formality and default values.

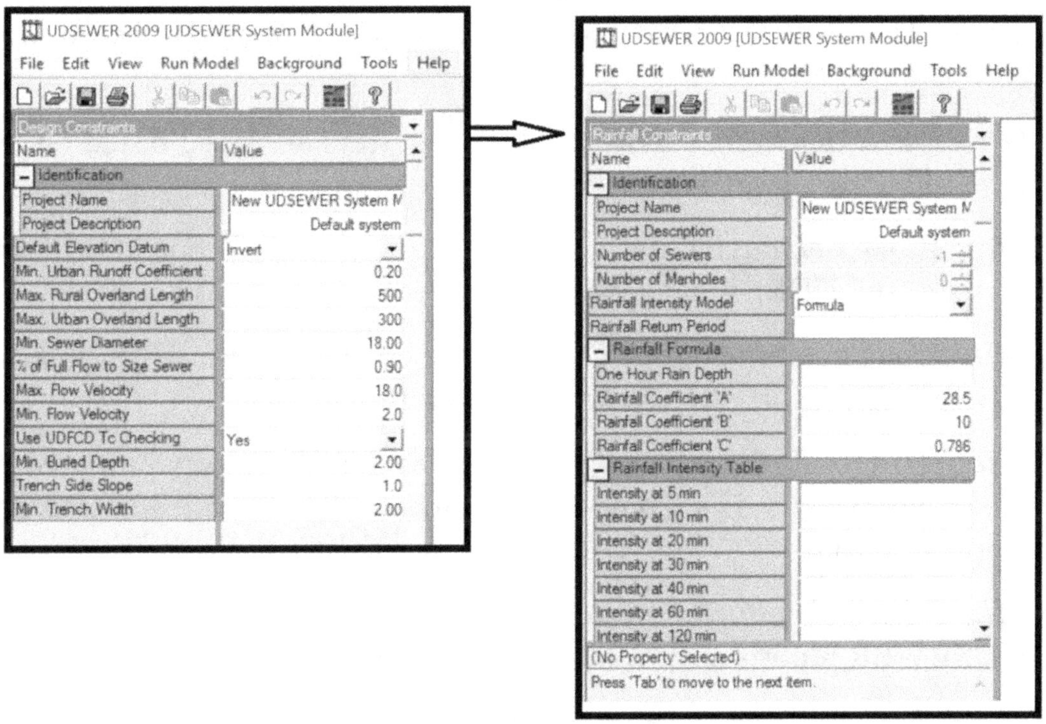

Figure 9.2. Inputs for Global Properties for UDSEWER.

The global properties can be typed in and will be applied to the entire manhole-sewer network. Under "*Design Constraints*", a project title and description are required; but they are not used in computations. Other variables are explained as:

1. ***Default Elevation Datum*:** Either sewer crown or invert elevation can be used to describe the sewer's vertical profile.
2. ***Min Urban Runoff Coefficient*:** a value of 0.2 is recommended. A catchment is considered urbanized if its runoff coefficient ≥Min Urban Runoff Coefficient.

3. ***Max Rural Overland Flow Length*:** a value of 500 (feet) is recommended. Beyond 500 feet, the overland flow becomes a swale (gutter) flow. When the catchment's runoff coefficient<0.2, a value >500 feet is NOT acceptable for overland flow length. UDSEWER will report such an inconsistency in input data.

4. ***Max Urban Overland Flow Length*:** a value of 300 (feet) is recommended. Beyond 300 feet, the overland flow becomes a swale (gutter) flow. When the catchment's runoff coefficient≥0.2, a value >300 feet is NOT acceptable for overland flow length. UDSEWER will report such an inconsistency in input data.

5. ***Min Sewer Diameter*:** a value of 18 (inches) shall be entered. When the hydraulically required diameter=14.8 inches, an 18-inch pipe is recommended for this new sewer. However the criterion of Min Sewer Diameter does not affect an existing sewer.

6. ***% to Full Flow to Size Sewer*:** All new sewers are sized based on 80 to 90% full in case of backwater effects.

7. ***Max Flow Velocity*:** Flow velocity in a sewer shall not be higher than 18 to 20 feet/second. UDSEWER reports all sewers with fast flows

8. ***Min Flow Velocity*:** Flow velocity in sewer shall be faster than 2 to 3 feet/sec. UDSEWER reports all sewers with slow flows.

9. ***Use UDFCD T_C Check*:** A "Yes" implies that T_d=min (Computed T_c, Regional T_c) or a "No" means T_d= Computed T_c, where T_d=rainfall duration and T_c= Time of Concentration.

10. ***Min Buried Depth*:** All sewer lines shall be covered with no less than 2-ft soils. UDSEWER reports any violation of min soil cover.

11. ***Trench Side Slope*:** A value of 0 to 1 shall be entered, depending on the soil stability.

12. ***Min Trench Width*:** A trench for sewer line shall be wider than two feet.

9.4.2. Input of Design Rainfall

The rainfall statistics can be entered in a formula form. Three parameters are required for an Intensity-Duration-Frequency (IDF) rainfall formula:

$$I(inch/hr) = \frac{k_1 P_1}{(k_2 + T_d)^{K_3}} \qquad (9.1)$$

Where I= rainfall intensity in inch/hr, T_d= duration in minutes, and k_1, k_2, *and* k_3 are input parameters. If the Option of "IDF table" is chosen, the user needs to enter the rainfall intensities for T_d= 5, 10, 30, 30, 40, 60 and 120 minutes. UDSEWER applies the linear interpolation and extrapolation between two duration points.

9.4.3. Manhole-Sewer Graphic Inputs

Each manhole is represented by an ID number, so is each sewer segment. The Manhole ID numbers in a system must not be duplicated; neither is any ID number among sewer ID numbers. At a manhole, UDSEWER can accept one outgoing sewer and up to four incoming sewers. On the right side of the screen in Figure 9.3, three icons: "*Pointer*", "*Pan*", *and* "*Manhole*" are available for the user to create a manhole-sewer system. A valid sewer system contains laterals coming into a trunk line. A trunk line must have one outfall manhole known as the "*system outfall*". To create an input file, use the icon: "*Manhole*" to create the system outfall element first. Assign a "zero" to be the identification number, "ID=0" which signifies this element to be the system outfall element. Later on, the tailwater input will be applied to the system-outfall element. The icon: "*Pointer*" allows the user to stretch or shrink the graphic display. The icon: "*Pan*" can move the window screen to display the sewer system.

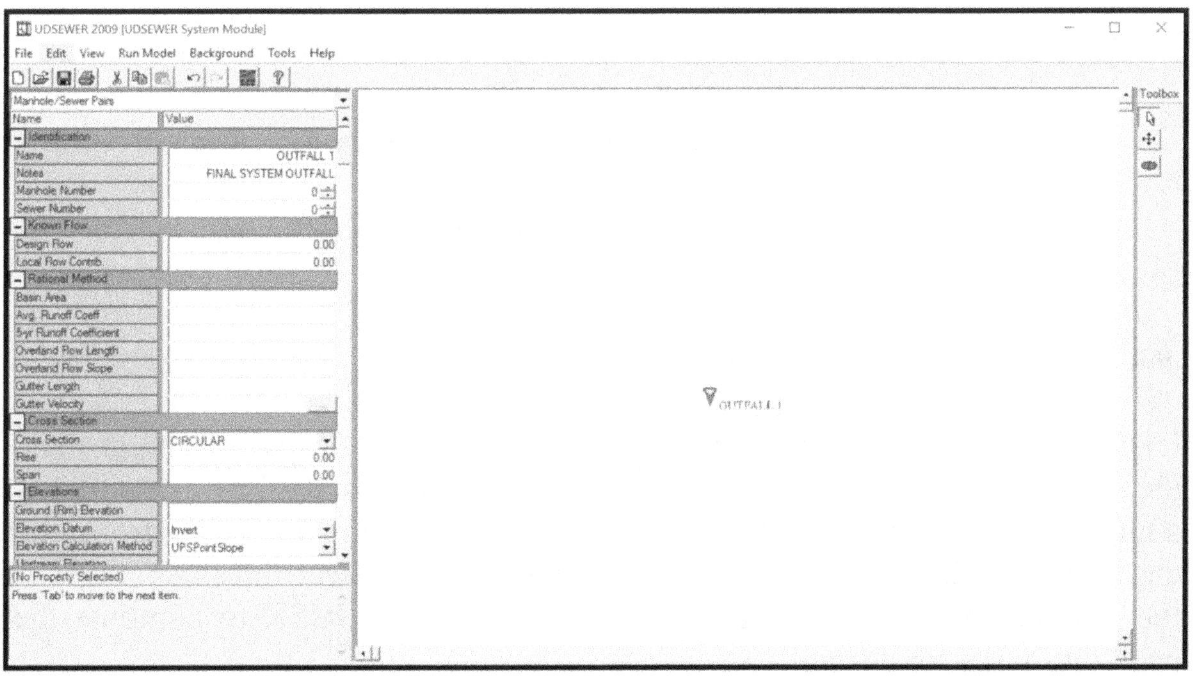

Figure 9.3. GUI for UDSEWER.

At this point, you are in the Graphic Utility Input (GUI) and ready for creating a new manhole-sewer layout. Next, you choose the *manhole icon* to drag and drop a sewer-manhole unit on the page till you build the sewer-manhole system for flow simulations. It is important to know that the program will apply the intelligence to identify the flow direction on a link (sewer line). Upon the completion of the sewer-manhole network as shown in Figure 9.4, *double clicks* on a manhole prompt the user to enter the data for the manhole, so does a sewer.

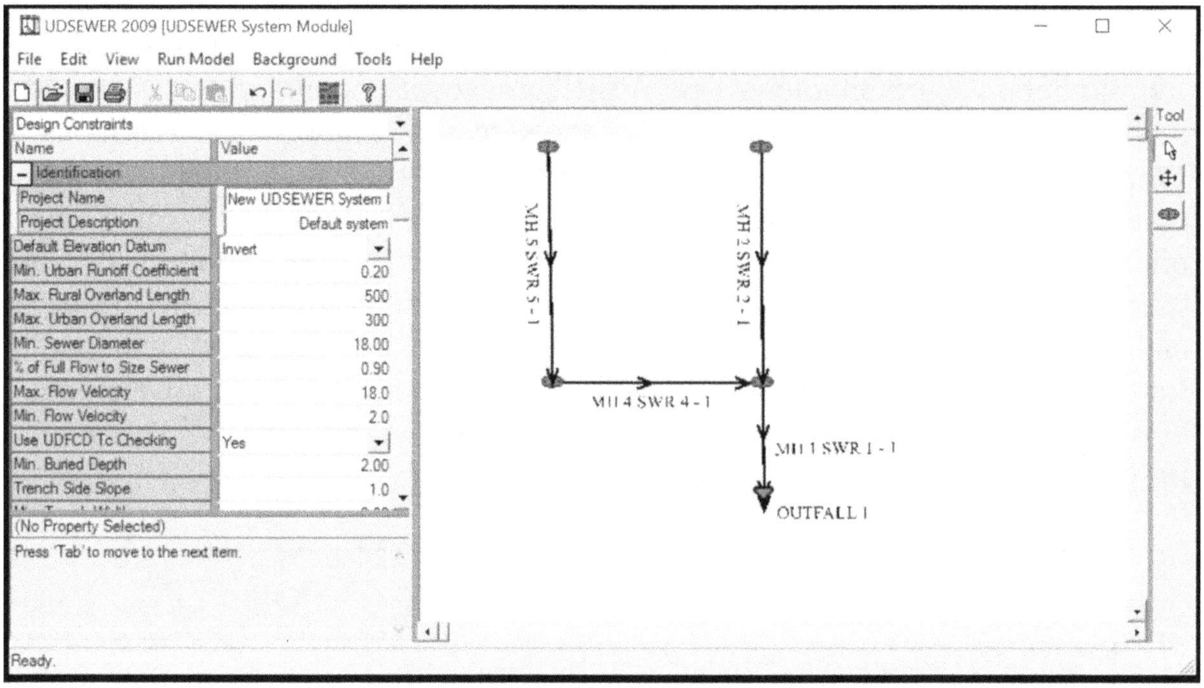

Figure 9.4. Creating a Manhole-Sewer System.

9.4.4. Manhole Input Data

UDSEWER applies the rational method to predict the design discharges and then sizes the sewers using Manning's equation. The design (or total) flow can only be changed at a manhole. At a manhole, the total flow includes: (1) the *local flow* from the local tributary area and (2) the *system flow* from the incoming sewer lines. Among all flow paths, the critical time of concentration is chosen by UDSEWER to compute the combined peak discharge for design.

There are 4 options for the user to define the local or total design flow at a manhole as:

(1) ***New Sewer Ready for Sizing***
The user must provide complete hydrologic data set to use the rational method for peak flow prediction, including basin area, 5-yr runoff coefficient, design runoff coefficient, overland flow and gutter flow parameters. UDSEWER will calculate the local and system flows, and then combine these two flows into the total discharge for sizing the outgoing sewer.

(2) ***Known Local Discharge from Local Tributary Area***
In this case, the local flow is already defined by other engineering report. The user must enter the known local flow, local tributary area and local runoff coefficient. The program will combine the local tributary area with the upstream tributary area together in order to determine the total flow to size the outgoing sewer.

(3) ***No Local Flow at Manhole***
In case that the manhole is a clean-out access that does not collect any local flow into the system. The user shall leave "blank" to all inputs for local flow. UDSEWER will combine all system flows from the incoming sewer lines through the manhole.

(4) ***Known Total Discharge***
When the outflow from a manhole is subject to a flow regulation such as pumps, the user may specify an allowable release rate in cfs. The input parameters include the total known flow, total cumulative tributary area upstream of this manhole, and area-weighted runoff coefficient. UDSEWER will reversely calculate the time of concentration for the specified total flow and check the flow continuity. UDSEWER reports the flow discontinuity between the inflow and outflow.

These options are useful when an off-site runoff is introduced into the sewer system under study or the flow is changed due to a pump operation at the manhole. The user can establish a new flow at the manhole to continue the flow simulation. When the user-defined known total flow is much different from the computed design discharge by the Rational method, the program will flag the difference as a warning.

When a large storm sewer system is being modeled, the user may divide the system into two smaller ones. The outflow from the upstream system can be transferred into the lower system by using the option of "*THE KNOWN TOTAL FLOW*".

9.4.5. Sewer Input Data

A sewer is described by *ground elevation, sewer profile, size, shape, bend loss coefficient, and lateral loss coefficient.* UDSEWER sizes the new sewers using circular pipes. Existing sewers are identified by the user defined known diameter for circular pipe, rise and span for elliptical pipe, or height and width for box pipe. Sewers with flat or negative slope must be treated as an existing sewer because the open channel hydraulics fail on a negative slope. The sewer vertical profile is defined by either the Option of "*UPS Point Slope*" using upstream crown/invert elevation with the length and slope, or the Option of "*DNS Point Slope*" using the downstream crown/invert elevation with the length and slope. The 3rd Option of "*Two Elevations*" is to use two point elevations and length of the sewer.

A bend loss in Figure 9.5 is entered to all sewers, based on the angle between incoming sewer line and outgoing sewer line. A lateral loss in Figure 9.4 is only entered to the sewer on the main line that has incoming laterals. Again, the lateral loss coefficient depends on the angle of the incoming lateral line and outgoing sewer line.

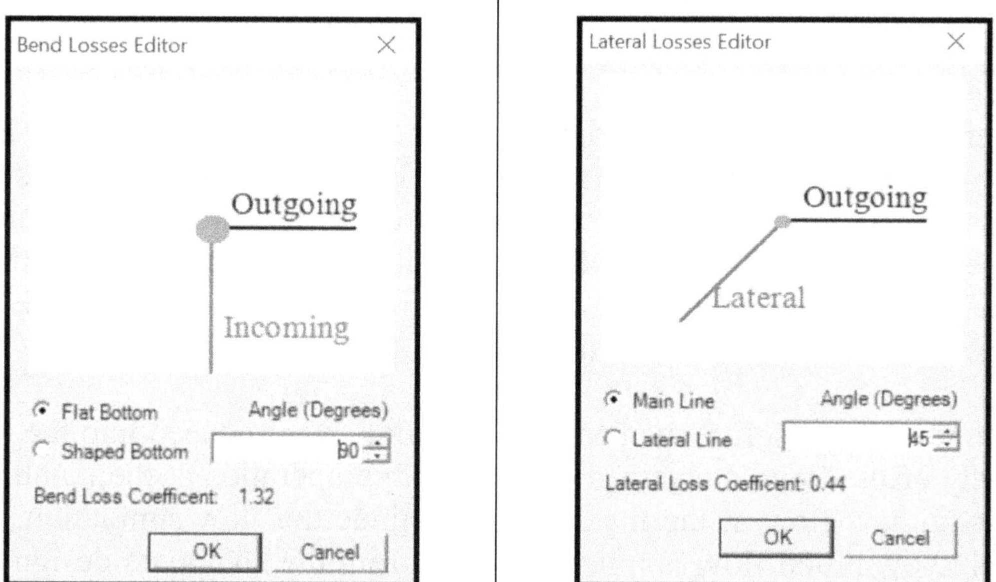

(a) bend loss (b) lateral loss

Figure 9.5. Bend and lateral Loss Coefficients.

The program does provide suggested commercial sewer sizes for both new and existing sewers, but the energy and hydraulic grade lines are computed using the existing sewers.

The program will apply open channel hydraulics, pressure flow theories, surcharge flow equations, and culvert hydraulics to analyze the flow conditions and energy losses.

9.5. EXECUTION OF INPUT FILE

9.5.1. Data Consistency Check

Upon completion of an input file, there is a tool in the UDSEWER program that can check and report any mistakes in the input file. Such a tool can be accessed by going to "*Tools Menu*", and selecting "*Validate the system*".

9.5.2. How to Run UDSEWER

Click "Run Model" on the pull down menu. Enter the tailwater elevation in feet at the system outfall. UDSEWER will produce an on-screen output file.

9.6. FILE MANAGEMENT

To save an input file, select the Option of "Save" under the file menu or choose "Save as" to use a new file name. The Option of "Print" under the file menu will print the output file.

UDSEWER can only read files created using UDSEWER file format. The file name created by UDSEWER has an extension: ".BSD" that means Binary Sewer Definition files. To load a network, select the Option of "Import" under the file menu.

9.7. GRAPHIC DISPLAY

Under "Tools" in the pull down menu, click "Draw Sewer Profile". You will see the dialogue box as shown in Figure 9.6. On the right side of the computer screen, you may click on "New" to set up a new sewer profile, "Edit" to revise an existing vertical profile, or "Remove" to delete an existing vertical profile.

To create a new profile, click "New". Next define the upstream "MH-SEWER" unit and downstream "MD-SEWER". Assign a name to this vertical profile. UDSEWER plots the sewer vertical profile with EGL and HGL as shown in Figure 9.7.

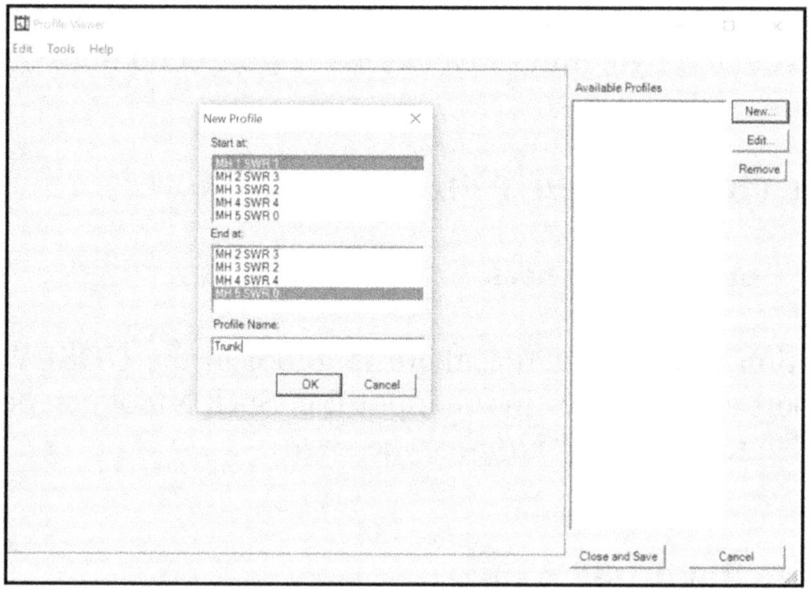

Figure 9.6. Dialogue Box to Define Sewer Profile.

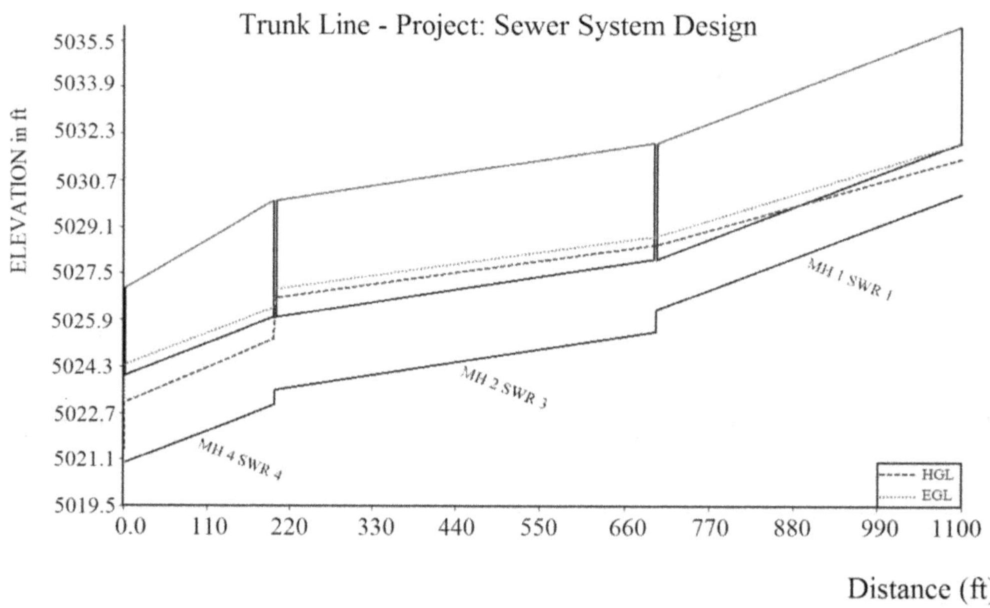

Figure 9.7. Example of Sewer Vertical Profile with HGL and EGL.

9.8. TROUBLESHOOTING

For troubleshooting go to the Troubleshooting Path Selector.

CHAPTER 10
STREET DRAINAGE PROJECT WORKSHOP ON UDSEWER

In this Chapter, we will lay out a street drainage system on a street map, and integrate street gutters, inlets, and sewers as a system to collect and convey the peaks flows predicted by the Rational method. The computer model, UDSEWER, will be introduced to design a sewer system under various design constrains and conditions.

10.1. PROJECT STUDY

This project is to determine the allowable street hydraulic conveyance capacity, to size the street inlets, and to layout storm sewer system to collect surface runoff generated from the tributary area. The sewer system is designed to drain storm runoff into a downstream lake.

The layout of streets, inlets, and sewers is illustrated in Figure 10.1. For this project, the dimensional units in Figure 10.1 are: X=______ (400 or 500), Y=_____(500 or 600) and Z=______ (30 or 44), depending on the assignment **from the instructor.**

For this project, inlets are installed at the street intersections. The storm sewers are laid to collect storm water from street intersections. At the outfall, the tailwater elevation is set at 5020 feet.

The design rainfall intensity formula (IDF) at the project area is given as:

$$I(inch/hr) = \frac{C_1 P_1}{(C_2 + T_d)^{C3}} \tag{10.1}$$

in which T_d = rainfall duration in minutes. The constants in IDF formula are: C_1 =28.5, C_2 =10, and C_3 = 0.786. The one-hr rainfall depths are given in Table 10.1.

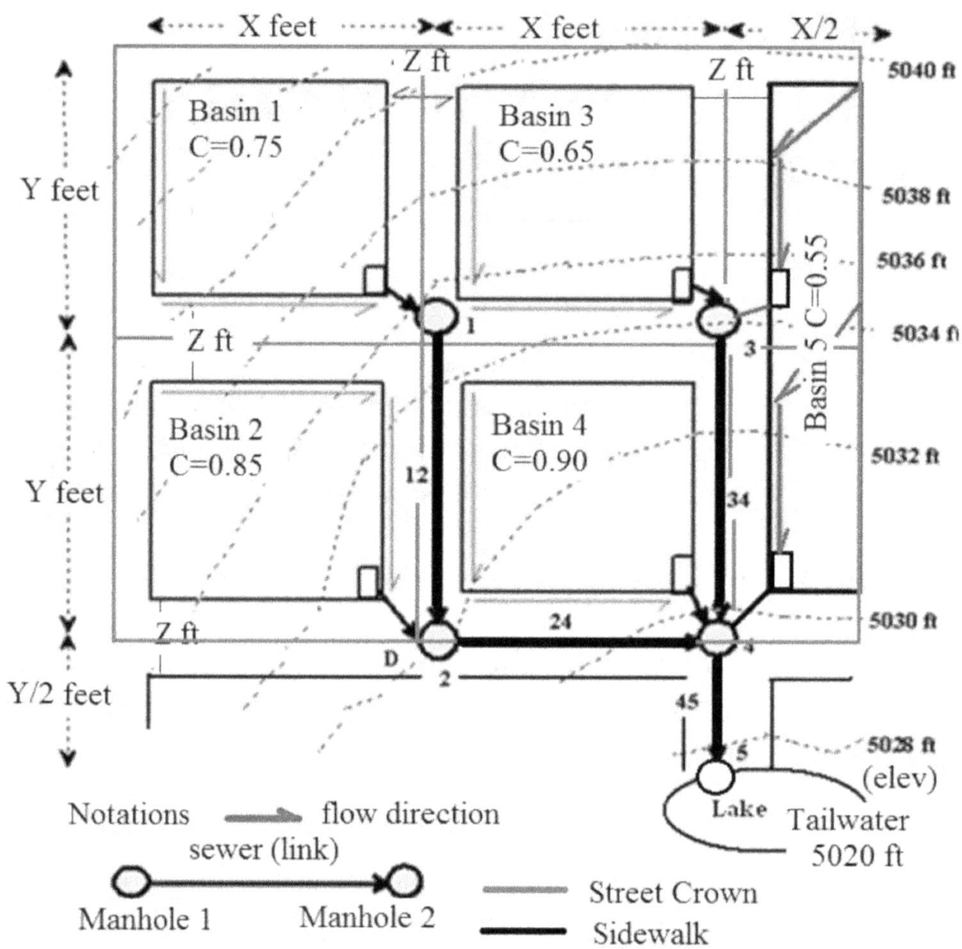

Figure 10.1. Layout of Design Project.

Table 10.1. One-hr Rainfall Depths.

One-hr Rain Depth	Return Period (yr)			
	2	5	10	100
P1 inch	0.95	1.35	1.61	2.60

For this project, the design return period is _____(5 or 100) years, **depending on the instructor.**

Task 1. Determine the street allowable hydraulic conveyance capacity $H_c=D_m=$ 6 inches, $D_s=2$ inches, $S_x=$ 0.02, and $S_o=$ 0.01 ft/ft or determined from the street map. As shown in Figure 10.2, Z= width of street and T=Z/2. If Z>40, set the width of emergency lane=5 feet or if Z<40, set the width of emergency lane= zero foot,

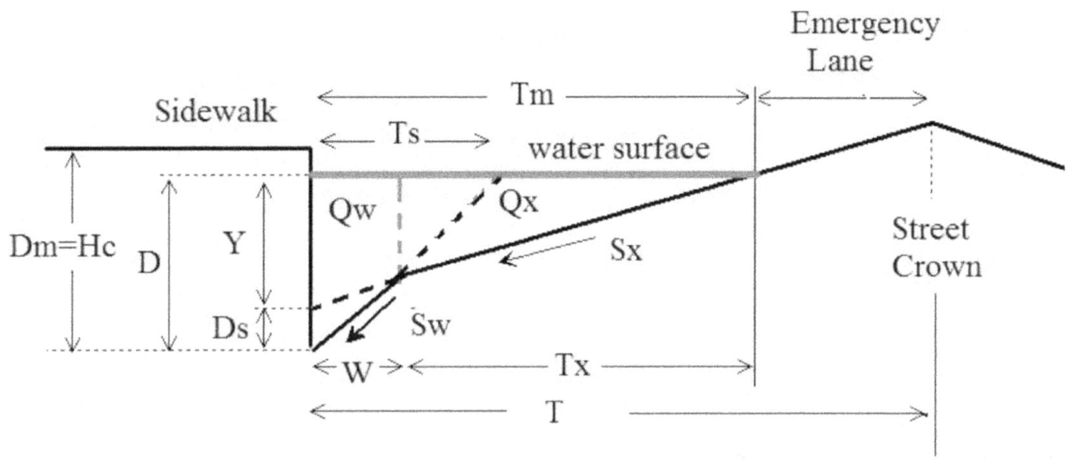

Figure 10.2. Street Cross for Project.

Tasks 2. Size the inlets.

Apply the Rational method to predict the design peak flows at the locations of inlets. Calculate the time of concentration based on the flow times. The overland flow time is calculated as:

$$T_o = \frac{0.395(1.1 - C_5)\sqrt{L_0}}{S_0^{\,0.33}} \tag{10.2}$$

in which T_o = overland flow time in minutes, C_5 = five-year runoff coefficient, L_o = overland flow length in feet (<300 feet), and S_o = overland slope in ft/ft. Both swale and gutter flow velocities can be estimated by the flow velocity equation as:

$$V = 2.0\sqrt{S} \text{ in which S = gutter slope in ft/ft.} \tag{10.3}$$

in which V= gutter flow velocity in ft/sec and S = gutter slope in ft/ft.

The computed time of concentration is the accumulation of flow times while the regional time of concentration is calculated as:

$$Tc_{regional}\,(min) = 10 + total\ flow\ length\ in\ feet/180 \tag{10.4}$$

(3.1) Determine the locations of inlets. An inlet is needed at a sump or wherever the peak flow on a grade exceeds the allowable street hydraulic conveyance capacity.

(3.2) Predict the peak flows at the proposed inlet locations. The design rainfall duration shall be the smaller one between the computed and the regional time of concentration.

(3.3) All inlets in sump shall be equipped with steel grates (Length x Width=2.95 x 1.85 feet). All inlets on grade shall be sized with curb-opening (Length = 5 feet). Consider clogging factor of 0.1 for curb-opening and 0.5 for grate. The debris amount at the site is subject to a decay factor of 0.25 when using multiple inlets. Estimate the number of grates or curb-opening inlets needed at each inlet location.

<u>TASK 3.</u> Design sewers and validate the sewer system with the EGL and HGL. The sewer system design constraints are:

(1) permissible flow velocity in a sewer between 2 and 18 feet/second,

(2) minimum earth coverage of 2 feet,

(3) minimum sewer diameter of 18 inches,

(4) minimum manhole drop of 0.50 foot,

(5) The outfall invert elevation is not lower than 5015 feet,

(6) The tailwater elevation at the downstream lake is 5020 feet, and

Figure 10.3 is the layout of the storm sewer system plotted by the computer model, UDSEWER.

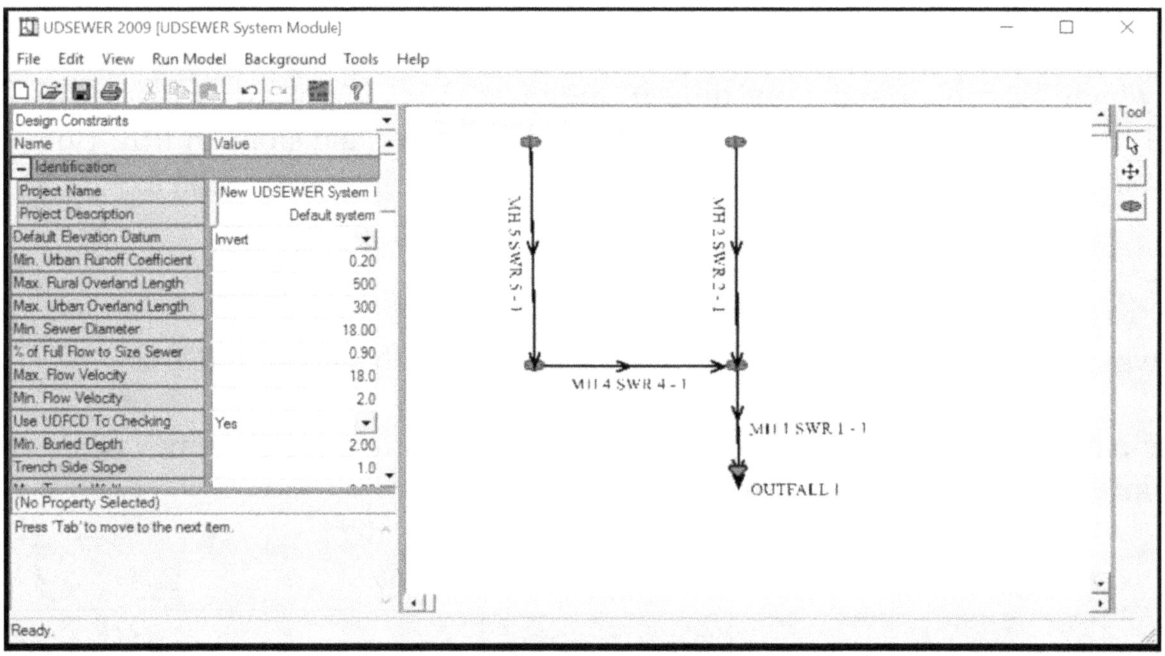

Figure 10.3. Graphic Utility Input (GUI) for UDSEWER.

For this case, the tailwater at the system exit is 5020 feet.

OUTPUT GENERTED FROM UDSEWER

Tailwater Elevation (ft): 5020.00

Manhole Input Summary:

		Given Flow		Sub Basin Information						
Element Name	Ground Elevation (ft)	Total Known Flow (cfs)	Local Contribution (cfs)	Drainage Area (Ac.)	Runoff Coefficient	5yr Coefficient	Overland Length (ft)	Overland Slope (%)	Gutter Length (ft)	Gutter Velocity (fps)
MH 5 SWR 0	5027.00	0.00	0.00	0.00	0.00	0.00	0.00	0.00	0.00	0.00
MH 4 SWR 4	5030.00	0.00	0.00	6.89	0.78	0.78	300.00	0.50	600.00	1.41
MH 2 SWR 3	5032.00	0.00	0.00	4.59	0.85	0.85	300.00	1.00	600.00	2.00
MH 1 SWR 1	5036.00	0.00	0.00	4.59	0.75	0.75	300.00	1.00	600.00	2.00
MH 3 SWR 2	5034.00	0.00	0.00	6.89	0.62	0.62	300.00	1.00	600.00	0.40

Manhole Output Summary:

	Local Contribution					Total Design Flow				
Element Name	Overland Time (min)	Gutter Time (min)	Basin Tc (min)	Intensity (in/hr)	Local Contrib (cfs)	Coeff. Area	Intensity (in/hr)	Manhole Tc (min)	Peak Flow (cfs)	Comment
MH 5 SWR 0	0.00	0.00	0.00	0.00	0.00	16.06	2.95	16.22	47.39	
MH 4 SWR 4	12.58	7.09	15.00	3.06	16.47	16.06	2.95	16.22	47.39	Used UDFCD Tc Maximum
MH 2 SWR 3	7.82	5.00	12.82	3.29	12.85	7.34	2.93	16.52	21.49	
MH 1 SWR 1	10.95	5.00	15.00	3.06	10.55	3.44	3.06	15.00	10.55	Used UDFCD Tc Maximum
MH 3 SWR 2	15.01	25.00	15.00	3.06	13.09	4.27	3.06	15.00	13.09	Used UDFCD Tc Maximum

Sewer Input Summary:

		Elevation			Loss Coefficients			Given Dimensions		
Element Name	Sewer Length (ft)	Downstream Invert (ft)	Slope (%)	Upstream Invert (ft)	Mannings n	Bend Loss	Lateral Loss	Cross Section	Rise (ft or in)	Span (ft or in)
MH 4 SWR 4	200.00	5021.00	1.00	5023.00	0.015	0.05	0.00	CIRCULAR	0.00 in	0.00 in
MH 2 SWR 3	500.00	5023.50	0.40	5025.50	0.015	0.05	0.25	CIRCULAR	0.00 in	0.00 in
MH 1 SWR 1	400.00	5026.25	1.00	5030.25	0.015	0.05	0.00	CIRCULAR	0.00 in	0.00 in
MH 3 SWR 2	400.00	5023.25	1.00	5027.25	0.015	0.05	0.00	CIRCULAR	0.00 in	0.00 in

Sewer Flow Summary:

	Full Flow Capacity		Critical Flow		Normal Flow						
Element Name	Flow (cfs)	Velocity (fps)	Depth (in)	Velocity (fps)	Depth (in)	Velocity (fps)	Froude Number	Flow Condition	Flow (cfs)	Surcharged Length (ft)	Comment
MH 4 SWR 4	57.96	8.20	26.90	8.37	24.75	9.15	1.18	Supercritical	47.39	0.00	
MH 2 SWR 3	22.54	4.59	18.91	6.59	23.41	5.23	0.65	Pressurized	21.49	500.00	
MH 1 SWR 1	13.77	5.72	14.52	5.95	13.77	6.31	1.11	Supercritical Jump	10.55	121.51	
MH 3 SWR 2	13.77	5.72	16.17	6.59	16.35	6.51	0.98	Pressurized	13.09	400.00	

Sewer Sizing Summary:

			Existing		Calculated		Used			
Element Name	**Peak Flow (cfs)**	**Cross Section**	**Rise**	**Span**	**Rise**	**Span**	**Rise**	**Span**	**Area (ft^2)**	**Comment**
MH 4 SWR 4	47.39	CIRCULAR	0.00 in	0.00 in	36.00 in	36.00 in	36.00 in	36.00 in	7.07	
MH 2 SWR 3	21.49	CIRCULAR	0.00 in	0.00 in	30.00 in	30.00 in	30.00 in	30.00 in	4.91	
MH 1 SWR 1	10.55	CIRCULAR	0.00 in	0.00 in	21.00 in	21.00 in	21.00 in	21.00 in	2.41	
MH 3 SWR 2	13.09	CIRCULAR	0.00 in	0.00 in	21.00 in	21.00 in	21.00 in	21.00 in	2.41	

Grade Line Summary:

Tailwater Elevation (ft): 5020.00

	Invert Elev.		Downstream Manhole Losses		HGL		EGL		
Element Name	**Downstream (ft)**	**Upstream (ft)**	**Bend Loss (ft)**	**Lateral Loss (ft)**	**Downstream (ft)**	**Upstream (ft)**	**Downstream (ft)**	**Friction Loss (ft)**	**Upstream (ft)**
MH 4 SWR 4	5021.00	5023.00	0.00	0.00	5023.06	5025.24	5024.36	1.97	5026.33
MH 2 SWR 3	5023.50	5025.50	0.01	0.62	5026.67	5028.49	5026.97	1.82	5028.78
MH 1 SWR 1	5026.25	5030.25	0.01	0.00	5028.50	5031.46	5028.80	3.21	5032.01
MH 3 SWR 2	5023.25	5027.25	0.02	0.00	5025.89	5029.51	5026.35	3.62	5029.97

Excavation Estimate:

The trench side slope is 1.0 ft/ft
The minimum trench width is 2.00 ft

					Downstream			Upstream				
Element Name	**Length (ft)**	**Wall (in)**	**Bedding (in)**	**Bottom Width (ft)**	**Top Width (ft)**	**Trench Depth (ft)**	**Cover (ft)**	**Top Width (ft)**	**Trench Depth (ft)**	**Cover (ft)**	**Volume (cu. yd)**	**Comment**
MH 4 SWR 4	200.00	4.00	6.00	6.67	10.00	6.83	2.67	12.00	7.83	3.67	398.77	
MH 2 SWR 3	500.00	3.50	6.00	6.08	11.50	7.29	3.71	11.50	7.29	3.71	957.27	
MH 1 SWR 1	400.00	2.75	4.00	5.21	10.75	6.31	3.77	10.75	6.31	3.77	600.82	
MH 3 SWR 2	400.00	2.75	4.00	5.21	12.75	7.31	4.77	12.75	7.31	4.77	774.89	

Total earth volume for sewer trenches = 2732 cubic yards.

10.2. CASE STUDY

Referring to Figure 10.4, an existing sewer system is subject to performance evaluation under the 100-year storm event. Equation 10.1 is applicable with P1=2.61 inches.

At the most upstream manhole, the design flow is 60 cfs generated from the tributary of 32.5 acres with a runoff coefficient of 0.5. At Mohole, M4, the pre-determined local flow of 40 cfs is added to the sewer system from a tributary area of 24 acres with C=05.

At M3, we will apply the Rational method to predict the peak flow from an area of 20 acres with C=0.5. At M4, there isn't any local flow to enter the system. At M5, the pump station will set the limit to the design flow at 120 cfs.

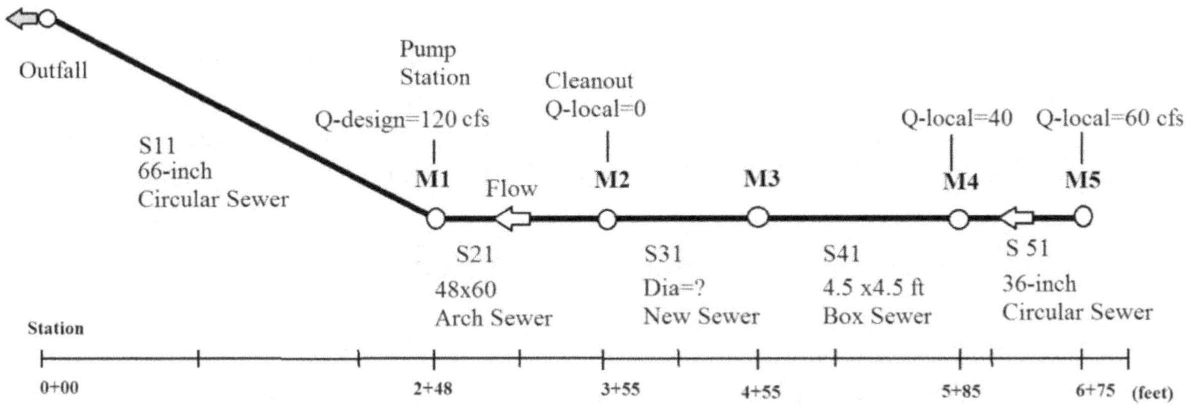

Figure 10.4. Sewer System Layout for Case Study.

Tables 10.2 and 10.3 specify the input parameters for the existing sewer system with their sizes, shapes and design flows.

Table 10.2. Manhole Information for Case Study.

Manhole ID Number	Ground Elevation feet	Tributary Area Acres	Runoff Coeff	Local Flow cfs	Design Flow Cfs	Overland Flow ft	Gutter Flow ft	Slope for Flow %
M5	206.00	32.5	0.5	60	0	0	0	0
M4	205.47	24	0.5	40	0	0	0	0
M3	204.49	20	0.5	0	0	300	1000	1.0
M2	203.14	0	0	0	0	0	0	0
M1	202.30	32.5+24+20	0.5	0	120	0	0	0
(outfall)	201.20	0						

Design flows are introduced into the sewer line from manholes. At a manhole, UDSEWER calculates the *local flow* from the local tributary areas and the *system flow* from the incoming sewer lines. The *design flow* is a combination of local flow and system flow. The local flow can be calculated by the Rational method for a new sewer or pre-determined as a known local flow. At a manhole such as cleanout, there isn't any local tributary area. UDSEWER will only maintain the continuity of the system flow through a manhole without a local flow. At a manhole, the outflow is controlled by a hydraulic structure such as pumps. The option of "*Known Design Flow*" shall be used with the user-defined flow. UDSEWER accepts the user-defined flow, and then reports the discontinuity between the system flow in sewer system and the flow defined by the user.

At M5 and M4 as listed in Table 10.2, the user must enter the local known flow, tributary area and runoff coefficient, and leave blank to the overland and gutter flow parameters. At M2, the manhole is a cleanout. Leave blank to all tributary and flow parameters because there isn't any local tributary area to M2. At M1, the outflow is limited to 120 cfs. The Option of "Known Design Flow" must be chosen. We need to enter the total cumulative area upstream of M1 and the area-weighted C

For all existing sewers, S51, S41, S21, and S11 in Table 10.3, the shapes and sizes must be entered. UDSEWER will accept the user-defined sewer sizes in all computations, and report the recommended sewer sizes for future improvements. Noted that both round and arch sewers are described by diameter or (rise x span) in inches while the box sewer is prescribed by it's (height x width) in feet. In this case, Sewer, S31, is new. UDSEWER will default a circular sewer with a proper size for S31.

Table 10.3. Sewer Information for Case Study.

Sewer ID	Length (feet)	Slope (percent)	Upstream Invert Elevation (ft)	Diameter (inches)	Height Or Rise (inches)	Width Or Span (inches)	Shape
51	90	1.15	200.00	36			Round
41	130	0.75	198.47		4.5 ft	4.5 ft	Box
31	100	0.85	196.99				New
21	107	1.25	195.64	60	48		Arch
11	248	0.65	193.80	66			Round

Figure 10.5 is the graphic input for this case study. It starts with the outfall manhole-sewer with ID=0. Next, drop and drag the Manhole Icon to add more manhole-sewer units.

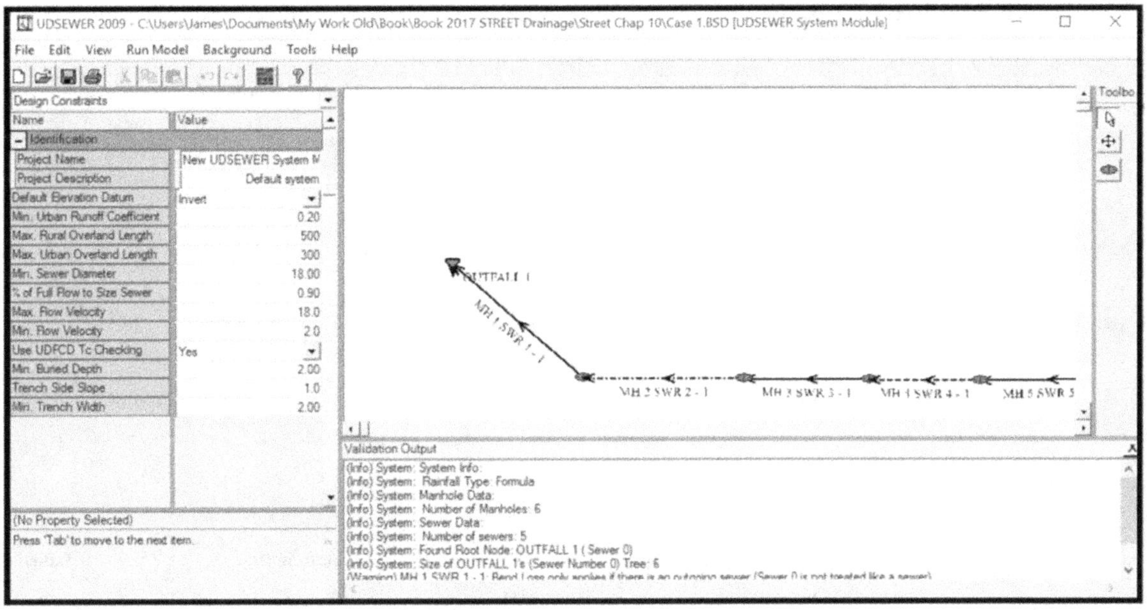

Figure 10.5. Graphic Input for Case Study.

Set the tailwater depth at elevation of 194 feet, UDSEWER produced the following outputs:

Selected Output from UDSEWER

Manhole Input Summary:

		Given Flow		Sub Basin Information						
Element Name	**Ground Elevation (ft)**	**Total Known Flow (cfs)**	**Local Contribution (cfs)**	**Drainage Area (Ac.)**	**Runoff Coefficient**	**5yr Coefficient**	**Overland Length (ft)**	**Overland Slope (%)**	**Gutter Length (ft)**	**Gutter Velocity (fps)**
OUTFALL 1	201.20	0.00	0.00	0.00	0.00	0.00	0.00	0.00	0.00	0.00
MH 1 SWR 1 - 1	202.30	120.00	0.00	76.50	0.50	0.50	0.00	0.00	0.00	0.00
MH 2 SWR 2 - 1	203.40	0.00	0.00	0.00	0.00	0.00	0.00	0.00	0.00	0.00
MH 3 SWR 3 - 1	204.49	0.00	0.00	20.00	0.50	0.50	300.00	1.00	1000.00	1.90
MH 4 SWR 4 - 1	205.47	0.00	40.00	24.00	0.50	0.50	0.00	0.00	0.00	0.00
MH 5 SWR 5 - 1	206.00	0.00	60.00	32.50	0.50	0.50	0.00	0.00	0.00	2.19

Manhole Output Summary:

	Local Contribution					Total Design Flow				
Element Name	Overland Time (min)	Gutter Time (min)	Basin Tc (min)	Intensity (in/hr)	Local Contrib (cfs)	Coeff. Area	Intensity (in/hr)	Manhole Tc (min)	Peak Flow (cfs)	Comment
OUTFALL 1	0.00	0.00	0.00	0.00	0.00	74.74	1.61	122.29	120.00	
MH 1 SWR 1 - 1	0.00	0.00	5.00	8.89	339.91	74.74	1.61	122.29	120.00	Used Minimum Tc
MH 2 SWR 2 - 1	0.00	0.00	0.00	0.00	0.00	36.49	3.65	36.48	133.30	
MH 3 SWR 3 - 1	18.76	8.77	17.22	5.56	55.63	36.49	3.65	36.48	133.30	Used UDFCD Tc Maximum
MH 4 SWR 4 - 1	0.00	0.00	42.23	3.33	40.00	26.49	3.68	36.03	97.51	Used Minimum Tc
MH 5 SWR 5 - 1	0.00	0.00	35.85	3.69	60.00	16.25	3.69	35.85	60.00	Used Minimum Tc

Sewer Input Summary:

		Elevation			Loss Coefficients			Given Dimensions		
Element Name	Sewer Length (ft)	Downstream Invert (ft)	Slope (%)	Upstream Invert (ft)	Mannings n	Bend Loss	Lateral Loss	Cross Section	Rise (ft or in)	Span (ft or in)
MH 1 SWR 1 - 1	248.00	192.06	0.7	193.80	0.014	0.03	0.00	CIRCULAR	66.00 in	66.00 in
MH 2 SWR 2 - 1	107.00	194.25	1.3	195.64	0.014	0.05	0.00	ELLIPSE	48.00 in	60.00 in
MH 3 SWR 3 - 1	100.00	196.09	0.9	196.99	0.014	0.05	0.00	CIRCULAR	0.00 in	0.00 in
MH 4 SWR 4 - 1	130.00	197.43	0.8	198.47	0.014	0.05	0.00	BOX	4.50 ft	4.50 ft
MH 5 SWR 5 - 1	90.00	198.92	1.2	200.00	0.014	0.05	0.00	CIRCULAR	36.00 in	36.00 in

Sewer Flow Summary:

	Full Flow Capacity		Critical Flow		Normal Flow						
Element Name	Flow (cfs)	Velocity (fps)	Depth (in)	Velocity (fps)	Depth (in)	Velocity (fps)	Froude Number	Flow Condition	Flow (cfs)	Surcharged Length (ft)	Comment
MH 1 SWR 1 - 1	261.59	11.01	36.48	8.91	31.38	10.77	1.33	Supercritical	120.00	0.00	
MH 2 SWR 2 - 1	221.25	14.81	40.76	10.35	31.35	13.92	1.67	Supercritical	133.30	0.00	
MH 3 SWR 3 - 1	173.70	10.92	40.76	10.35	35.46	12.04	1.32	Supercritical	133.30	0.00	
MH 4 SWR 4 - 1	208.51	10.30	29.32	8.87	25.68	10.12	1.22	Supercritical	97.51	0.00	
MH 5 SWR 5 - 1	68.03	9.62	30.02	9.53	26.26	10.86	1.33	Supercritical	60.00	0.00	

Sewer Sizing Summary:

			Existing		Calculated		Used			
Element Name	Peak Flow (cfs)	Cross Section	Rise	Span	Rise	Span	Rise	Span	Area (ft^2)	Comment
MH 1 SWR 1 - 1	120.00	CIRCULAR	66.00 in	66.00 in	54.00 in	54.00 in	66.00 in	66.00 in	23.76	
MH 2 SWR 2 - 1	133.30	ELLIPSE	48.00 in	60.00 in	48.00 in	48.00 in	48.00 in	60.00 in	14.94	
MH 3 SWR 3 - 1	133.30	CIRCULAR	0.00 in	0.00 in	54.00 in	54.00 in	54.00 in	54.00 in	15.90	
MH 4 SWR 4 - 1	97.51	BOX	4.50 ft	4.50 ft	4.50 ft	4.50 ft	4.50 ft	4.50 ft	20.25	
MH 5 SWR 5 - 1	60.00	CIRCULAR	36.00 in	36.00 in	36.00 in	36.00 in	36.00 in	36.00 in	7.07	

On the top menu, under the Option: "TOOLS", click on "Draw Sewer Profile". The sewer vertical profile and HGL and EGL will be plotted after specifying the sewer line with the beginning and end manholes. The user may save, edit, or remove the profile files. Figure 10.6 is a sewer profile for the case study.

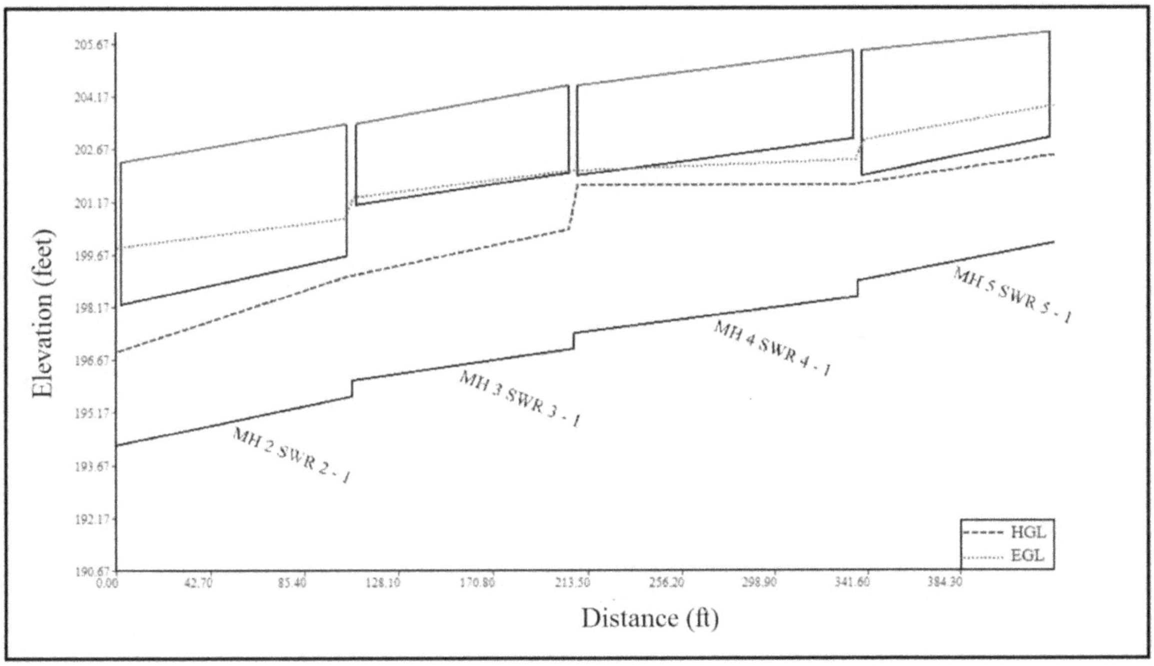

Figure 10.6. UDSEWER Plot of Sewer Profile and HGL and EGL.

10.3. HOMEWORK

Figure HW 10.1 shows the layout of street inlets and sewers. This system has five basins with its own runoff coefficient. Sewers are designed to collect storm water and to drain the 5-year event into Cherry Creek that has a 5-year water surface elevation at 5001 feet. The rainfall intensity, I, is represented as:

$$I(inch/hr) = \frac{40}{(10+T_d)^{0.786}}$$ in which T_d = rainfall duration in minutes.

(1) determine the street allowable hydraulic capacity

width of street= 44 feet, width of emergency lane= 5 feet

curb height= 6 inches, gutter depression = 2 inches

(2) place inlets

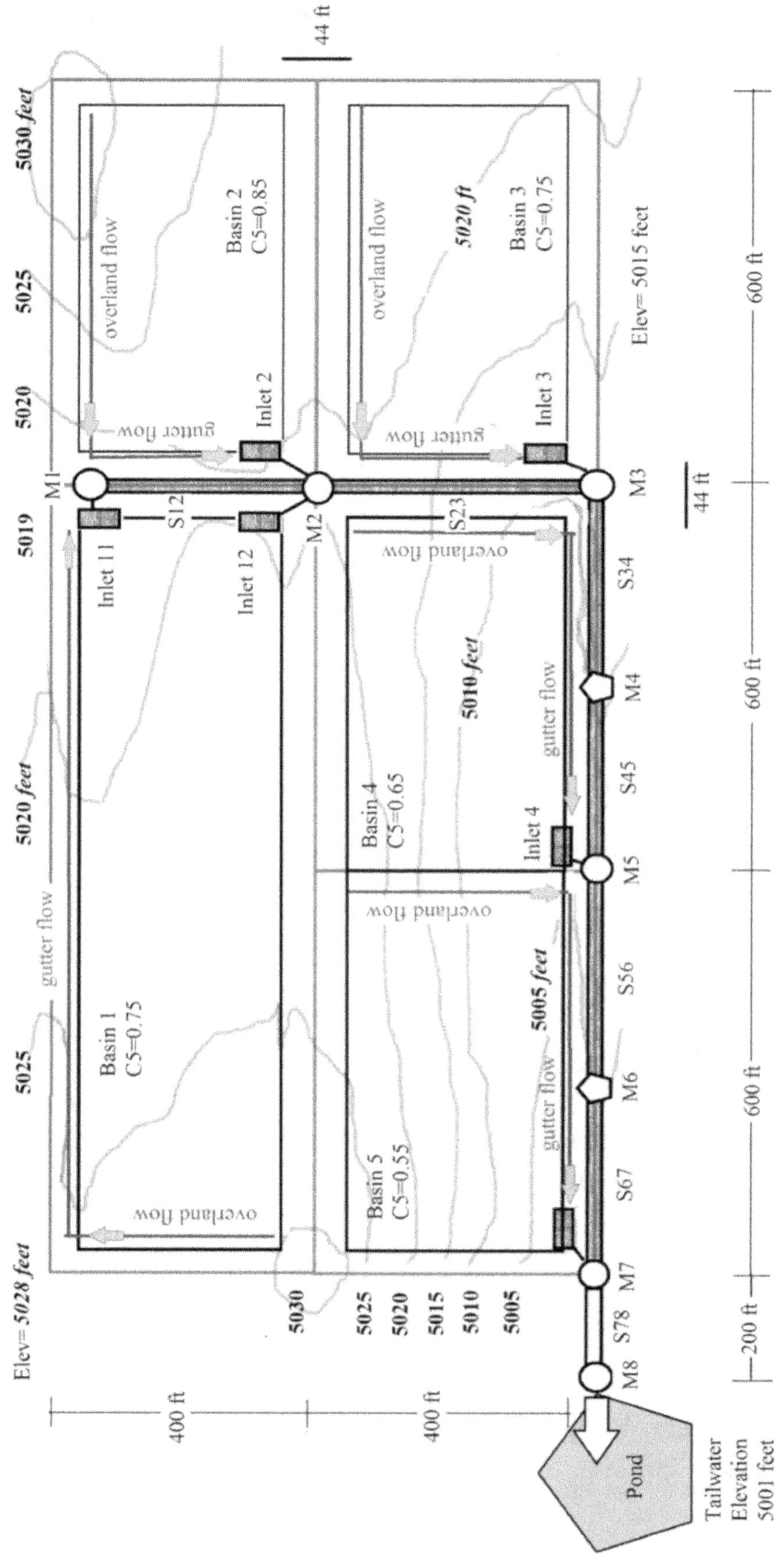

Figure HW 10.1. Layout of Sewer and Inlet Systems.

(3) size the inlets using grates for sump and curb-opening for on-grade
(4) verify that the given pipe sizes are adequate to pass the given design flow,
(5) plot the sewer vertical profile with EGL and HGL.

The above tasks are subject to the following design constraints:

a. The minimum velocity in sewers is 2.0 fps.
b. The maximum velocity in sewers is 18.0 fps
c. The minimum soil cover along sewers is 2.0 ft.
d. The maximum length between manholes is 500 feet.
e. The maximum drop at a manhole is 5 feet.
f. The minimum drop at a manhole is 0.25 ft.

www.ingramcontent.com/pod-product-compliance
Lightning Source LLC
LaVergne TN
LVHW082342200926
848481LV00012B/3422